PROCEEDINGS OF THE

FIRST INTERNATIONAL CONFERENCE ON

GENETIC ALGORITHMS

AND THEIR APPLICATIONS

July 24-26, 1985
at the
Carnegie-Mellon University
Pittsburgh, PA

Sponsored By

Texas Instruments, Inc.

Naval Research Laboratory

John J. Grefenstette
Editor

 Psychology Press
Taylor & Francis Group

New York London

First Published by

Lawrence Erlbaum Associates, Inc., Publishers
10 Industrial Avenue
Mahwah, New Jersey 07430

Transferred to Digital Printing 2009 by Psychology Press
270 Madison Ave, New York NY 10016
27 Church Road, Hove, East Sussex, BN3 2FA

ISBN 0-8058-0426-9

Publisher's Note
The publisher has gone to great lengths to ensure the quality of this
reprint but points out that some imperfections in the original may be apparent.

PREFACE

It has been ten years since the publication of John Holland's seminal book, *Adaptation in Natural and Artificial Systems*. One of the major contributions of this book was the formulation of a class of algorithms, now known as *Genetic Algorithms (GA's)*. which incorporates metaphors from natural population genetics into artificial adaptive systems. Since the publication of Holland's book, interest in GA's has spread from the University of Michigan to research centers throughout the U.S., Canada, and Great Britain. GA's have been applied to a striking variety of areas, from machine learning to image processing to combinatorial optimization. The great range of application attests to the power and generality of the underlying approach. However, much of the GA research has been reported only in Ph. D. theses and informal workshops. This Conference was organized to provide a forum in which the diverse groups involved in GA research can share results and ideas concerning this exciting area.

ACKNOWLEDGEMENTS

On behalf of the Conference Committee, it is my pleasure to acknowledge the support of our sponsors: Texas Instruments, Inc., and the Navy Center for Applied Research in Artificial Intelligence at the Naval Research Laboratory. Special thanks go to Dave Davis for arranging the support from TI and to Steve Smith for handling the local arrangements.

John J. Grefenstette
Program Chair

Conference Committee

John H. Holland	University of Michigan (*Conference Chair*)
Lashon B. Booker	Navy Center for Applied Research in AI
Kenneth A. De Jong	George Mason University
John J. Grefenstette	Vanderbilt University (*Program Chair*)
Stephen F. Smith	Carnegie-Mellon Robotics Institute (*Local Arrangements*)

TABLE OF CONTENTS

SESSION 5

SESSION 6

FRIDAY, JULY 26, 1985

SESSION 7

PROPERTIES OF THE BUCKET BRIGADE ALGORITHM

John H. Holland

The University of Michigan

The bucket brigade algorithm is designed to solve the apportionment of credit problem for massively parallel, message-passing, rule-based systems. The apportionment of credit problem was recognized and explored in one of the earliest significant works in machine learning (Samuel [1959]). In the context of rule-based systems it is the problem of deciding which of a set of early acting rules should receive credit for "setting the stage" for later, overtly successful actions. In the systems of interest here, in which rules conform to the standard condition/action paradigm, a rule's overall usefulness to the system is indicated by a parameter called its *strength*. Each time a rule is active, the bucket brigade algorithm modifies the strength so that it provides a better estimate of the rule's usefulness in the contexts in which it is activated.

The bucket brigade algorithm functions by introducing an element of competition into the process of deciding which rules are activated. Normally, for a parallel message-passing system, all rules having condition parts satisfied by some of the messages posted at a given time are automatically activated at that time. However, under the bucket brigade algorithm only some of the satisfied rules are activated. Each satisfied rule makes a *bid*, based in part on its strength, and only the highest bidders become active (thereby posting the messages specified by their action parts). The size of the bid depends upon both the rule's strength *and* the specificity of the rule's conditions. (The rule's specificity is used on the broad assumption that, other things being equal, the more information required by a rule's conditions, the more likely it is to be "relevant" to the particular situation confronting it). In a specific version of the algorithm used for classifier systems, the bid of classifier C at time t is given by

$$b(C,t) = cr(C)s(C,t),$$

where $r(C)$ is the specificity of rule C (equal, for classifier systems, to the difference between the total number of defining positions in the condition and the number of "don't cares" in the condition), $s(C,t)$ is the strength of the rule at time t, and c is a constant considerably less than 1

1

(e.g., 1/4 or 1/8).

The essence of the bucket brigade algorithm is its treatment of each rule as a kind of mid-level entrepreneur (a "middleman") in a complex enconomy. When a rule C wins the competition at time t, it must <u>decrease</u> its strength by the amount of the bid. Thus its strength on time-step t+1, after winning the competition, is given by

$$s(C, t+1) = S(C, t) - b(C, t) = (1 - cr(C))S(C, t).$$

In effect C has paid for the privilege of posting its message. Moreover this amount is actually paid to the classifers that sent messages satisfying C's conditions -- in the simplest formulation the bid is split equally amongst them. These message senders are C's *suppliers*, and each receives its share of the payment from the *consumer* C. Thus, if C_1 has posted a message that satisfies one of C's conditions, C_1 has its strength increased so that

$$s(C_1, t+1) = S(C_1, t) + b(C, t)/n(C, t) = (1 - cr(C)/n(C,t))S(C,t),$$

where $n(C, t)$ is the number of classifiers sending messages that satisfy C at time t.

In terms of the economic metaphor, the suppliers $\{C_1\}$ are paid for setting up a situation usable by consumer C. C, on the next time step, changes from consumer to supplier because it has posted its message. If other classifiers then bid because they are satisfied by C's message, and if they win the bidding competition, then C in turn will receive some fraction of those bids. C's survival in the system depends upon its turning a profit as an intermediary in these local transactions. In other words, when C is activated, the bid it pays to its suppliers must be less (or, at least, no more) than the average of the sum of the payments it receives from its consumers.

It is important that this process involves no complicated "bookkeeping" or memory over long sequences of action. When activated, C simply pays out its bid on one time-step, and is immediately paid by its consumers (if any) on the next time-step. The only variation on this transaction occurs on time-steps when there is payoff from the environment. Then, all classifiers active on that time-step receive equal fractions of the payoff in addition to any payments from classifiers active on the next time-step. In effect, the environment is the system's ultimate consumer. From a global point of view, a given classifier C is likely to be

2

profitable only if its usual consumers are profitable. The profitability of any chain of consumers thus depends upon their relevance to the ultimate consumer. Stated more directly, the profitability of a classifier depends upon its being coupled into sequences leading to payoff.

As a way of illustrating the bucket brigade algorithm, consider a set of 2-condition classifiers where, for each classifier, condition 1 attends to messages from the environment and condition 2 attends to messages from other classifiers in the set. As above, let a given classifier C have a bid fraction b(C) and strength s(C,t) at time t. Note that condition 1 of C defines an equivalence class E in the environment consisting of those environmental states producing messages satisfying the condition.

Consider now the special case where the activation of C produces a response r that transforms states in E to states in another equivalence class E' having an (expected) payoff u. Under the bucket brigade algorithm, when C wins the competition under these circumstances its strength will change from s(C,t) to

$$s(C,t+1) = s(C,t) - b(C)s(C,t) + u$$
$$+ \text{(any bids C receives from classifiers active on the next time-step)}.$$

Assuming the strength of C is small enough that its bid b(C)s(C,t) is considerably less than u, the usual case for a new rule or for a rule that has only been activated a few times, the effect of the payoff is a considerable strengthening of rule C.

This strengthening of C has two effects. First, C becomes more likely to win future competitions when its conditions are satisfied. Second, rules that send messages satisfying one (or more) of C's conditions will receive higher bids under the bucket brigade, because b(C)s(C,t+1) > b(C)s(C,t).

Both of these effects strongly influence the development of the system. The increased strength of C means that response r will be made more often to states in E when C competes with other classifiers that produce different responses. If states in E' are the only payoff states accessible from E, and r is the only response that will produce the required transformation from states in E to states in E', then the higher probability of a win for C translates into a higher payoff rate to the classifier system.

3

Of equal importance, C's higher bids mean that rules sending messages satisfying C's second condition will be additionally strengthened because of C's higher bids. Consider, for example, a classifier C_0 that transforms environmental states in some class E_0 to states in class E by evoking response r_0. That is, C_0 acts upon a causal relation in the environment to "set the stage" for C. If C_0 also sends a message that satisfies C's second condition, then C_0 will benefit from the "stage setting" because C's higher bid is passed to it via the bucket brigade.

It is instructive to contrast the "stage setting" case with the case where some classifier, say C_1, sends a message that satisfies C but *does not* transform states in E_1 (the environmental equivalence class defined by its first condition) to states in E. That is, C_1 attempts to "parasitize" C, extracting bids from C via the bucket brigade without modifying the environment in ways suitable for C's action. Because C_1 is not instrumental in transforming states in E_1 to states in E, it will often happen that activation of C_1 is not followed by activation of C on the subsequent time-step because C's first (environmental) condition is not satisfied. Every time C_1 is activated without a subsequent activation of C it suffers a loss because it has paid out its bid $b(C_1)s(C_1,t)$, without receiving any income from C. Eventually C_1's strength will decrease to the point that it is no longer a competitor. (There is a more interesting case where C_0 and C_1 manage to become active simultaneously, but that goes beyond the confines of the present illustration).

One of the most important consequences of the bidding process is the automatic emergence of default hierarchies in response to complex environments. For rule-based systems a "default" rule has two basic properties:
   1) It is a general rule with relatively few specified properties and many "don't cares" in its condition part, and
   2) when it wins a competition it is often in error, but it still manages to profit often enough to survive.
It is clear that a default rule is preferable to no rule at all, but, because it is often in error, it can be improved. One of the simplest improvements is the addition of an "exception" rule that responds to situations that cause

the default rule to be in error. Note that, in attempting to identify the error-causing situations, the condition of the exception rule specifies a *subset* of the set of messages that satisfy the default rule. That is, the condition part of the exception rule *refines* the condition part of the default rule by using *additional* identifying bits (properties). Because rule discovery algorithms readily generate and test refinements of existing strong rules, useful exception rules are soon added to the system.

As a direct result of the bidding competition, an exception rule, once in place, actually aids the survival of its parent default rule. Consider the case where the default rule and the exception rule attempt to set a given effector to a different values. In the typical classifier system this conflict is resolved by letting the highest bidding rule set the effector. Because the exception rule is more specific than the default rule, and hence makes a higher bid, it usually wins this competition. In winning, the exception rule actually prevents the default rule from paying its bid. This outcome saves the the default rule from a loss, because the usual effect of an error, under the bucket brigade, is activation of consumers that do not bid enough to return a profit to the default rule. In effect the exception protects the default from some errors. Similar arguments apply, under the bucket brigade algorithm, when the default and the exception only influence the setting of effectors indirectly through intervening, coupled classifiers.

Of course the exception rules may be imperfect themselves, selecting some error-causing cases, but making errors in other cases. Under such circumstances, the exception rules become default rules relative to more detailed exceptions. Iteration of the above process yields an ever more refined, and efficient, default hierarchy. The process improves both overall performance and the profitability of each of the rules in the hierarchy. It also uses fewer rules than would be required if all the rules were developed at the most detailed level of the hierarchy (see Holland, Holyoak, Nisbett, and Thagard [1986]). The bucket brigade algorithm strongly encourages the top-down discovery and development of such hierarchies (cf. Goldberg [1983] for a concrete example).

At first sight, consideration of long sequences of coupled rules would seem to uncover an important limitation of the bucket brigade algorithm. Because of its local nature, the bucket brigade algorithm can only propagate strength back along a chain of suppliers through repeated activations of the whole sequence. That is, on the first repetition of a sequence leading to payoff, the increment in strength is propagated to the immediate precursors of the payoff rule(s). On the second repetition it is propagated to the precursors of the precursors, etc. Accordingly, it takes on the order of n repetitions of the sequence to propagate the increments back to rules that "set the stage" n steps before the final payoff. However, this observation is misleading because certain kinds of rule can serve to "bridge" long sequences.

The simplest "bridging action" occurs when a given rule remains active over, say, T successive time-steps. Such a rule passes increments back over an interval of T time-steps on the *next* repetition of the sequence. This qualification takes on importance when we think of a rule that shows persistent activity over an *epoch* -- an interval of time characterized by a broad plan or activity that the system is attempting to execute. For the activity to be persistent, the condition of the epoch-marking rule must be general enough to be satisfied by just those properties or cues that characterize the epoch. Such a rule, if strong, marks the epoch by remaining active for its duration.

To extract the consequences of this persistent activation, consider a concrete plan involving a sequence of activities, such as a "going home" plan. The sequence of coupled rules used to execute this plan on a given day will depend upon variable requirements such as "where the car is parked", "what errands have to be run", etc. These detailed variations will call upon various combinations of rules in the system's repertoire, but the epoch-marking "going home" rule D will be active throughout the execution of each variant. In particular, it will be active both at the beginning of the epoch and at the time of payoff at the end of the plan ("arrival home"). As such it "bridges" the whole epoch.

Consider now a rule I that initiates the plan and is coupled to (sends a message satisfying) the general epoch-marking rule D. The *first repetition* of the sequence initiated by I will result in the strength of I being incremented. This comes about because D is strengthened by being active at the time of payoff and, because it is a consumer of I's message, it passes this increment on to I the very next time I is activated. D "supports" I as an element of the "going home" plan. The result is a kind of one-shot learning in which the earliest elements in a plan are rewarded on the very next use. This occurs despite the local nature of the bucket brigade algorithm. It requires only the presence of a general rule -- a kind of default -- that is activated when some general kind of activity or goal

6

is to be attained. An appropriate rule discovery algorithm, such as a genetic algorithm, will soon couple more detailed rules to the epoch-marking rule. And, much as in the generation of a default hierarchy, these detailed rules can give rise to further refined offspring. The result is an emergent plan hierarchy going from a high-level sketch through progressive refinements yielding ways of combining progressively more detailed components (rule clusters) to meet the particular constraints posed by the current state of the environment. In this way a limited repertoire of rules can be combined in a variety of ways, and in parallel, to meet the perpetual novelty of the environment.

References.

Goldberg, D. E. *Computer-aided Gas Pipeline Operation Using Genetic Algorithms and Machine Learning*. Ph. D. Dissertation (Civil Engineering). The University of Michigan. 1983.

Holland, J. H., Holyoak, K. J,, Nisbett, R. E., and Thagard, P. R. *Induction: Learning, Discovery, and the Growth of Knowledge*. [forthcoming, MIT Press].

Samuel, A. L. "Some studies in machine learning using the game of checkers." *IBM Journal of Research and Development, 3*. 211-232, 1959.

GENETIC ALGORITHMS AND RULE LEARNING
IN
DYNAMIC SYSTEM CONTROL

David E. Goldberg
Department of Engineering Mechanics
The University of Alabama

ABSTRACT

In this paper, recent research results [1] are presented which demonstrate the effectiveness of genetic algorithms in the control of dynamic systems. Genetic algorithms are search algorithms based upon the mechanics of natural genetics. They combine a survival-of-the-fittest among string structures with a structured, yet randomized, information exchange to form a search algorithm with some of the innovative flair of human search. While randomized, genetic algorithms are no simple random walk. They efficiently exploit historical information to speculate on new search points with improved performance.

Two applications of genetic algorithms are considered. In the first, a tripartite genetic algorithm is applied to a parameter optimization problem, the optimization of a serial natural gas pipeline with 10 compressor stations. While solvable by other methods (dynamic programming, gradient search, etc.) the problem is interesting as a straightforward engineering application of genetic algorithms. Furthermore, a surprisingly small number of function evaluations are required (relative to the size of the discretized search space) to achieve near-optimal performance.

In the second application, a genetic algorithm is used as the fundamental learning algorithm in a more complete rule learning system called a learning classifier system. The learning system combines a complete string rule and message system, an apportionment of credit algorithm modeled after a competitive service economy, and a genetic algorithm to form a system which continually evaluates its present rules while forming new, possibly better, rules from the bits and pieces of the old. In an application to the control of a natural gas pipeline, the learning system is trained to control the pipeline under normal winter and summer conditions. It is also trained to detect the presence or absence of a leak with increasing accuracy.

INTRODUCTION

Many industrial tasks and machines that once required human intervention have been all but completely automated. Where once a person tooled a part, a machine tools, senses, and tools again. Where once a person controlled a machine, a computer controls, senses, and continues its task. Repetitive tasks requiring a high degree of precision have been most susceptible to these extreme forms of automated control. Yet despite these successes, there are still many tasks and mechanisms that require the attention of a human operator. Piloting an airplane, controlling a pipeline, driving a car, and fixing a machine are just a few examples of ordinary tasks which have resisted a high degree of automation. What is it about these tasks that has prevented more autonomous, automated control? Primarily, each of the example tasks requires, not just a single capability, but a broad range of skills for successful performance. Furthermore, each task requires performance under circumstances which have never been encountered before. For example, a pilot must take off, navigate, control speed and direction, operate auxiliary equipment, communicate with tower control, and land the aircraft. He may be called upon to do any or all of these tasks under extreme weather conditions or with equipment malfunctions he has never faced before. Clearly, the breadth and perpetual novelty of the piloting task (and similarly complex task environments) prevents the ordinary algorithmic solution used in more repetitive chores. In other words, difficult environments are difficult because not every possible outcome can be anticipated in advance, nor can every possible response be predefined. This truth places a premium on adaptation.

In this paper, we attack some of these issues by examining research results in two distinct, but related problems. In the first, the steady state control of a serial gas pipeline is optimized using a genetic algorithm. While the optimization problem itself is unremarkable (a straightforward parameter optimization problem which has been solved by other methods), the genetic algorithm approach we adopt is noteworthy because it draws from the most successful and longest lived search algorithm known to man (natural genetics + survival-of-thefittest). Furthermore, the GA approach is provably efficient in its exploitation of important similarities, and thus connects to our own notions of innovative or creative search. In the second problem, we use a genetic algorithm as a primary discovery mechanism in a larger rule learning system called a learning classifier system (LCS). In this particular application the LCS learns to control a simulated natural gas pipeline. Starting from a random rule set the LCS learns appropriate rules for high performance control under normal summer and winter conditions; additionally it learns to detect simulated leaks with increasing accuracy.

A TRIPARTITE GENETIC ALGORITHM

Genetic algorithms are different from the normal search methods encountered in engineering optimization in the following ways:

1. GA's work with a coding of the parameter set not the parameters themselves.

2. GA's search from a population of points.

3. GA's use probabilistic not deterministic transition rules.

Genetic algorithms require the natural parameter set of the optimization problem to be coded as a finite length string. A variety of coding schemes can and have been used successfully. Because GAs work directly with the underlying code they are difficult to fool because they are not dependent upon continuity of the parameter space and derivative existence.

In many optimization methods, we move gingerly from a single point in the decision space to the next using some decision rule to tell us how to get to the next point. This point-by-point method is dangerous because it often locates false peaks in multimodal search spaces. GA's work from a database of points simultaneously (a population of strings) climbing many peaks in parallel, thus reducing the probability of finding a false peak.

Unlike many methods, GAs use probabilistic decision rules to guide their search. The use of probability does not suggest that the method is simply a random search, however. Genetic algorithms are quite rapid in locating improved performance.

For our work, we may consider the strings in our population of strings to be expressed in a binary alphabet containing the characters {0,1}. Each string is of length ℓ and the population contains a total of n such strings. Of course, each string may be decoded to a set of physical parameters according to our design. Additionally, we assume that with each string (parameter set) we may evaluate a fitness value. Fitness is defined as the non-negative figure of merit we are maximizing. Thus, the fitness in genetic algorithm work corresponds to the objective function in normal optimization work.

A simple genetic algorithm which gives good results is composed of three operators:

1. Reproduction

2. Crossover

3. Mutation

With our simple genetic algorithm we view reproduction as a process by which individual strings are copied according to their fitness. Highly fit strings receive higher numbers of copies in the mating pool. There are many ways to do this; we simply give a proportionately higher probability of reproduction to those strings with higher fitness (objective function value). Reproduction is thus the survival-of-the-fittest or emphasis step of the genetic algorithm. The best strings make more copies for mating than the worst.

After reproduction, simple crossover may proceed in two steps. First, members of the newly reproduced strings in the mating pool are mated at random. Second, each pair of strings undergoes crossing over as follows: an integer position k along the string is selected uniformly at random on the interval $(1, \ell-1)$. Two new strings are created by swapping all characters between positions 1 and k inclusively.

For example, consider two strings A and B of length 7 mated at random from the mating pool created by previous reproduction:

A = a1 a2 a3 a4 a5 a6 a7
B = b1 b2 b3 b4 b5 b6 b7

Suppose the roll of a die turns up a four. The resulting crossover yields two new strings A' and B' following the partial exchange

A' = b1 b2 b3 b4 a5 a6 a7
B' = a1 a2 a3 a4 b5 b6 b7

The mechanics of the reproduction and crossover operators are surprisingly simple, involving nothing more complex than string copies and partial string exchanges; however, together the emphasis step of reproduction and the structured, though randomized, information exchange of crossover give genetic algorithms much of their power. At first this seems surprising. How can such simple (computationally trivial) operators result in anything useful let along a rapid and relatively robust search mechanism? Furthermore, doesn't it seem a little strange that chance should play such a fundamental role in a directed search process? The answer to the second question was well recognized by the mathematician J. Hadamard [2]:

> We shall see a little later that the possibility of imputing discovery to pure chance is already excluded....On the contrary, that there is an intervention of chance but also a necessary work of unconsciousness, the latter implying and not contradicting the former.... Indeed, it is obvious that invention or discovery, be it in mathematics or anywhere else, takes place by combining ideas.

The suggestion here is that while discovery is not a result of pure chance, it is almost certainly guided by directed serendipity. Furthermore, Hadamard hints that a proper role for chance is to cause the juxtaposition of different notions. It is interesting that genetic algorithms adopt Hadamard's mix of direction and chance in a manner which efficiently builds new solutions from the best partial solutions of previous trials.

To see this, consider a population of n strings over some appropriate alphabet coded so that each is a complete IDEA or prescription for performing a particular task (in our coming example, each string is a description of how to operate all 10 compressors on a natural gas pipeline.). Substrings within each string (IDEA) contain various NOTIONS of what's important or relevant to the task. Viewed in this way, the population contains not just a sample of n IDEAS, rather it contains a multitude of NOTIONS and rankings of those NOTIONS for task performance. Genetic algorithms carefully exploit this wealth of information about important NOTIONS by 1) reproducing quality NOTIONS according to their performance and 2) crossing these NOTIONS with many other high performance NOTIONS from other strings. Thus, the act of crossover with previous reproduction speculates on new IDEAS constructed from the high performance building blocks (NOTIONS) of past trials.

If reproduction according to fitness combined with crossover give genetic algorithms the bulk of their processing power, what then is the purpose of the mutation operator? Not surprisingly there is much confusion about the role of mutation in genetics (both natural and artificial). Perhaps it is the result of too many B movies detailing the exploits of mutant eggplants that devour portions of Chicago, but whatever the cause for the confusion, we find that mutation plays a decidedly secondary role in the operation of genetic algorithms. Mutation is needed because, even though reproduction and crossover effectively search and recombine extant NOTIONS, occasionally they may become overzealous and lose some potentially useful genetic material (1's or 0's at particular locations). The mutation operator protects against such an unrecoverable loss. In the simple tripartite GA, mutation is the occasional random alteration of a string position. In a binary code, this simply means changing a 1 to a 0 and vice versa. By itself, mutation is a random walk through the string space. When used sparingly with reproduction and crossover it is an insurance policy against premature loss of important NOTIONS.

That the mutation operator plays a secondary role we simply note that the frequency of mutation to obtain good results in empirical genetic algorithm studies is on the order of 1 mutation per thousand bit (position) transfers. Mutation rates are similarly small in natural populations which leads us to conclude that mutation is appropriately considered as a secondary mechanism.

The underlying processing power of genetic algorithms is understood in more rigorous terms by considering the notion of a NOTION more carefully. If two or more strings (IDEAS) contain the same NOTION there are similarities between the strings at one or more positions. To consider the number and form of the possible relevant similarities we consider a schema [3] or similarity template; a similarity template is simply a string over our original alphabet {1,0} with the addition of a wild card or don't care character *. For example, with string length ℓ = 7 the schema 1*0**** represents all strings with a 1 in the first position and a 0 in the third position. A simple counting argument shows that while there are only 2^ℓ strings, there are 3^ℓ well-defined schemata or possible templates of similarity. Furthermore, it is easy to show that a particular string is itself a representative of 2^ℓ different schemata. Why is this interesting? The interesting part comes from considering the effect of reproduction and crossover on the multitude of schemata contained in a population of n strings (at most $n \cdot 2^\ell$ schemata). Reproduction on average gives exponentially more samples to the observed best similarity patterns (a near-optimal sampling strategy if we consider a multi-armed bandit problem). Second, crossover, combines schemata from different strings so that only very long defining length schemata (relative to the string length) are interrupted. Thus, short defining length schemata are propagated generation to generation by giving exponentially increasing samples to the observed best, and all this goes on in parallel with little explicit book-keeping or special memory other than the population of n strings. How many of the $n \cdot 2^\ell$ schemata are usefully processed per generation? Using a conservative estimate, Holland has shown that $O(n^3)$ schemata are usefully sampled per generation. This compares favorably with the number of function evaluations (n), and because this processing leverage is so important (and apparently unique to genetic algorithms) Holland gives it a special name, implicit parallelism. In the next section we exploit this leverage in the optimization of a natural gas pipeline.

THE TRIPARTITE GENETIC ALGORITHM IN NATURAL GAS PIPELINE OPTIMIZATION

We apply the genetic algorithm to the steady state serial natural gas pipeline problem of Wong and Larson [4]. As mentioned previously, the problem is not remarkable. Wong and Larson successfully used a dynamic programming approach and gradient procedures have also been used. Our goal here is to connect with extant optimization and control literature. We also look at some of the issues we face in applying genetic algorithms to more difficult problems where standard techniques may be inappropriate.

We envision a serial system with an alternating sequence of 10 compressors and 10 pipelines. A fixed pressure source exists at the inlet; gas is delivered at line pressure to the delivery point. Along the way, compressors boost pressure using fuel taken from the line. Modeling relationships for the steady flow of an ideal gas are well studied. We adopt Wong and Larson's formulation for consistency. The reader interested in more modeling detail should refer to their original work.

Along with the usual modeling relationships, we must pose a reasonable objective function and constraints. For this problem, we use Wong and Larson's objective function and constraint specification. Specifically, we minimize the summed horsepower over the 10 compressor stations in the serial line subject to maximum and minimum pressure constraints as well as maximum and minimum pressure ratio constraints. Constraints in these state variables are adjoined to the problem using an exterior penalty method. Whenever a constraint is violated a penalty cost is added to the objective function in proportion to the square of the violation. As we shall see in a moment, constraints in control variables may be handled with the choice of some appropriate finite coding.

As discussed in the previous section, one of the necessary conditions for using a genetic algorithm is the ability to code the underlying parameter set as a finite length string. This is no real limitation as every user of a digital computer or calculator knows; however, there is motivation for constructing special, relatively crude codings. In this study, the full string is formed from the concatenation of 10, four bit substrings where each substring is a mapped fixed point binary integer (precision = 1 part in 16) representing the difference in squared pressure across each of the ten compressor stations. This rather crude discretization gives an average precision in pressure of 34 psi over the operating range 500-1000 psia.

The model, objective function, constraints, and genetic algorithm have been programmed in Pascal. We examine results from a number of independent trials and compare to published results. To initiate simulation, a starting population of 50 strings is selected at random. For each trial of the genetic algorithm we run to generation 60. This represents a total of 50*61=3050 function evaluations per independent trial. The results from three trials are shown in Figure 1. This figure shows the cost of the best string of each generation as the solution proceeds. At first, performance is poor. After sufficient

genetic action, near-optimal results are obtained. In all three cases, near-optimal results are obtained by generation 20 (1050 function evaluations).

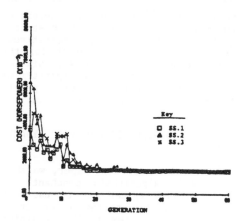

Figure 1.   Best-of-Generation
Results - Steady
Serial Problem

To better understand these results, we compare the best solution obtained in the first trial (run SS.1) to the optimal results obtained by dynamic programming. A pressure profile is presented in Figure 2. The GA results are very close to the dynamic programming solution, with most of the difference explained by the large discretization errors associated with the GA solution.

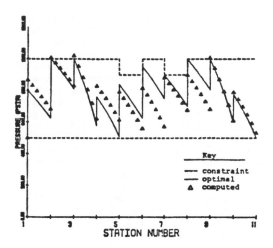

Figure 2.   Pressure Profile -
Run SS.1 Steady
Serial Problem

To gain a feel for the search rapidity of the genetic algorithm, we must compare the number of points searched to the size of the search space. Recall that in this problem, near-optimal performance is obtained after only 1050 function evaluations. To put this in perspective, with a string of length 40, there are 2^ℓ different possible solutions in the search space ($2^{40} = 1.1\text{e}12$). Therefore, we obtain near-optimal results after searching only 1e-7% of the possible alternatives. If we were, for example, to search for the best person among the worlds 4.5 billion people as rapidly as the genetic algorithm we would only need to talk to 4 or 5 people before making our near-optimal selection.

A LEARNING CLASSIFIER SYSTEM FOR DYNAMIC SYSTEM CONTROL

In the remainder of this paper, we show how the genetic algorithm's penchant for discovery in string spaces may be usefully applied to search for string rules in a learning classifier system (LCS). Learning classifier systems are the latest outgrowth of Holland's continuing work on adaptive systems [5]. Others have continued and extended this work in a variety of areas ranging from visual pattern recognition to draw poker [6-8].

A learning classifier system (LCS) is an artificial system that learns rules, called classifiers, to guide its interaction in an arbitrary environment. It consists of three main elements:

1. Rule and Message System

2. Apportionment of Credit System

3. Genetic Algorithm

A schematic of an LCS is shown in Figure 3. In this schematic, we see that the rule and message system receives environmental information through its sensors, called detectors, which decode to some standard message format. This environmental message is placed on a message list along with a finite number of other internal messages generated from the previous cycle. Messages on the message list may activate classifiers, rules in the classifier store If activated a classifier may then be chosen to send a message to the message list for the next cycle. Additionally, certain messages may call for external action through a number of action triggers called effectors. In this way, the rule and message system combines both external and internal data to guide behavior and the state of mind in the next state cycle.

In an LCS, it is important to maintain simple syntax in the primary units of information, messages and classifiers. In the current study messages are ℓ-bit (binary) strings and classifiers are 3ℓ-position strings over the alphabet $\{0,1,\#\}$. In this alphabet the # is a wild card, matching a 0 or a 1 in a given message. Thus, we maintain powerful pattern recognition capability with simple structures.

ENVIRONMENT

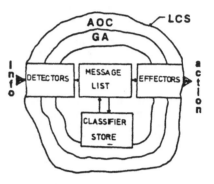

Figure 3. Schematic - Learning Classifier System

In traditional rule-based expert systems, the value or rating of a rule relative to other rules is fixed by the programmer in conjunction with the expert or group of experts being emulated. In a rule learning system, we don't have this luxury. The relative value of different rules is one of the key pieces of information which must be learned. To facilitate this type of learning, Holland has suggested that rules coexist in a competitive service economy. A competition is held among classifiers where the right to answer relevant messages goes to the highest bidders with this payment serving as a source of income to previously successful message senders. In this way, a chain of middlemen is formed from manufacturer (source message) to message consumer (environmental action and payoff). The competitive nature of the economy insures that the good rules survive and that bad rules die off.

In addition to rating existing rules, we must also have a way of discovering new, possibly better, rules. This, of course, is the appropriate role for our genetic algorithm. In the learning classifier system application, we must be less cavalier about replacing entire string populations each generation, and we should pay more attention to the replacement of low performers by new strings; however, the genetic algorithm adopted in the LCS is very similar to the simple tripartite algorithm described earlier.

Taken together, the learning classifier system with a computationally complete and convenient rule and message system, an apportionment of credit system modeled after a competitive service economy, and the innovative search of a genetic algorithm, provides a unified framework for investigating the learning control of dynamic systems. In the next section we examine the application of an LCS to natural gas pipeline operation and leak detection.

A LEARNING CLASSIFIER SYSTEM CONTROLS A PIPELINE

A pipeline model, load schedule, and upset conditions are programmed and interfaced to the LCS. We briefly discuss this environmental model and present results of normal operations and upset tests.

A model of a pipeline has been developed which accounts for linepack accumulation and frictional resistance. User demand varies on a daily basis and depends upon the weather. Different patterns may be used for winter and summer operation. In addition to normal summer and winter conditions, the pipeline may be subjected to a leak upset. During any given time step, a leak may occur with a specified leak probability. If a leak occurs, the leak flow, a specified value, is extracted from the upstream junction and persists for a specified number of time steps.

The LCS receives a message about the pipeline condition every time step. A template for that message is shown in Figure 4. The system has complete, albeit imperfect and discrete, knowledge of its state including inflow, outflow, inlet pressure, outlet pressure, pressure rate change, season, time of day, time of year, and current temperature reading.

In the pipeline task, the LCS has a number of alternatives for actions it may take. It may send out a flow rate chosen from one of four values, and it may send a message indicating whether a leak is suspected or not.

The LCS receives reward from its trainer depending upon the quality of its action in relation to the current state of the pipeline. To make the trainer evervigilant, a computer subroutine has been written which administers the reward consistently. This is not a necessary step, and reward can come from a human trainer.

Under normal operating conditions we examine the performance of the learning classifier system with and without the genetic algorithm enabled. Without the genetic algorithm, the system is forced to make do with its original set of rules. The results of a normal operating test are presented in Figure 5. Both runs with the LCS outperform a random walk (through the operating alternatives). Furthermore, the run with genetic algorithm enabled is superior to the run without GA. In this figure, we show time-averaged total evaluation versus time of simulation (maximum reward per timestep = 6).

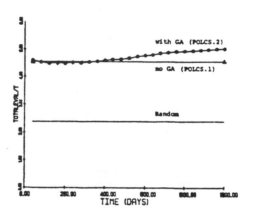

Figure 5. Time-averaged TOTALEVAL vs. Time. Normal Operations. Runs POLCS.1 & POLCS.2

More dramatic performance differences are noted when we have the possibility of leaks on the system. Figure 6 shows the time-averaged total evaluation versus time for several runs with leak upsets. Once again the LCS is initialized with random rules and permitted to learn from external reward. Both LCS runs outperform the random walk and the run with GA clearly beats the run with no new rule learning. To understand this, we take a look at some auxiliary performance measures. In Figure 7 we see

| PI | QI | PO | QO | DP | TOD |TT|TP| TAG |

Variable	Description	min	max	# of positions
PI	inlet pressure	0	2000	2
QI	inlet flow	0	80	2
PO	outlet pressure	0	2000	2
QO	outlet flow	0	80	2
DP	u. s. pressure rate	-200	200	2
TOD	time of day	0	24	2
TT	time of year	0	1	1
TP	temperature	0	1	1

Figure 4. Pipeline LCS Environmental Message Template

the percentage of leaks alarmed correctly versus time. Strangely, the run without GA alarms a higher percentage of leaks than the run with GA. This may seem counterintuitive until we examine the false alarm statistics in Figure 8. The run without GA is only able to alarm a high percentage of leaks correctly because it has so many false alarms. The run with GA decreases its false alarm percentage, while increasing its leaks correct percentage.

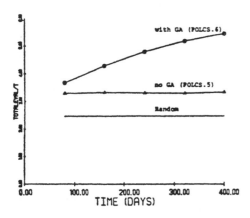

Figure 6. Time-averaged TOTALEVAL vs. Time - Leak Runs - POLCS.5 & POLCS.6

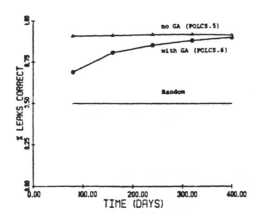

Figure 7. Percentage of Leaks Correct vs. Time Runs POLCS.5 & POLCS.6

CONCLUSIONS

In this paper, we examined the performance of a genetic algorithm in two applications. In the first, a tripartite genetic algorithm consisting of reproduction, crossover, and mutation was applied to the

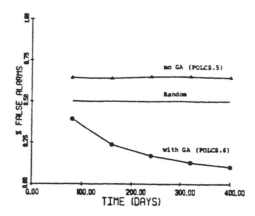

Figure 8. Percentage of False Alarms vs. Time Runs POLCS.5 & POLCS.6

optimization of a natural gas pipeline's operation. The control space was coded as 40 bit binary strings. Three initial populations of 50 strings were chosen at random. The genetic algorithm was started and in all three cases, very near-optimal performance was obtained after only 20 generations (1050 function evaluations).

In the second application, a genetic algorithm was the primary discovery mechanism in a larger rule-learning system called a learning classifier system. The LCS, consisting of a syntactically simply rule and message system, an apportionment of credit mechanism based on a competitive service economy, and a genetic algorithm, was taught to operate a gas pipeline under winter and summer conditions. It also was trained to alarm correctly for leaks while minimizing the number of false alarms.

REFERENCES

1. Goldberg, D. E., "Computer-Aided Pipeline Operation using Genetic Algorithms and Rule Learning," Ph.D. Dissertation, University of Michigan, Ann Arbor, 1983.

2. Hadamard, J., The Psychology of Invention in the Mathematical Field, Princeton University Press, Princeton, 1945.

3. Holland, J. H., Adaptation in Natural and Artificial Systems, University of Michigan Press, Ann Arbor, 1975.

4. Wong, P. J. and R. E. Larson, "Optimization of Natural Gas Pipeline Systems via Dynamic Programming," IEEE Trans. Auto. Control, vol. AC-13, no. 5, pp. 475-481, October, 1968.

5. Holland, J. H. and J. S. Reitman, "Cognitive Systems Based on Adaptive Algorithms," in *Pattern-Directed Inference Systems*, Waterman, D. A. and F. Hayes-Roth (eds.), pp. 313-329, Academic Press, New York, 1978.

6. Smith, S. F., "A Learning System Based on Genetic Adaptive Algorithms," Ph.D. dissertation, University of Pittsburgh, Pittsburgh, 1980.

7. Booker, L. B., "Intelligent Behavior as an Adaptation to the Task Environment," Ph.D. dissertation, University of Michigan, Ann Arbor, 1982.

8. Wilson, S., "Adaptive 'Cortical' Pattern Recognition," unpublished manuscript, Rowland Institute of Science, Cambridge, MA, 1983.

KNOWLEDGE GROWTH IN AN ARTIFICIAL ANIMAL

by

Stewart W. Wilson

Rowland Institute for Science, Cambridge MA 02142

ABSTRACT

Results are presented of experiments with a simple artificial animal model acting in a simulated environment containing food and other objects. Procedures within the model that lead to improved performance and perceptual generalization are discussed. The model is designed in the light of an explicit definition of intelligence which appears to apply to all animal life. It is suggested that study of artificial animal models of increasing complexity would contribute to understanding of natural and artificial intelligence.

INTRODUCTION

The science of understanding and realizing intelligence in artificial systems needs a definition of intelligence. Every science needs good definitions of the problems it addresses. But in the artificial intelligence field there has been a hesitancy about defining intelligence. For example, on the first page of a recent, widely used AI textbook we find: "A definition in the usual sense seems impossible because intelligence appears to be an amalgam of so many information-representation and information-processing talents."[1] For many AI goals, this omission is not important. But the lack of a good working definition can lead to uncertainty in evaluating progress toward understanding intelligence *per se*, even though results are in other respects substantial.

This paper reports work using an artificial, behaving, animal model to study intelligence at a primitive level. An explicit definition of intelligence is adopted, and guides construction of the model. The definition has intuitive appeal and apparent applicability to the range of life from human beings to very primitive animals. Because of this range, some results with the primitive animal model should provide insight into intelligence in general.

A DEFINITION OF INTELLIGENCE

A good definition should be relatively simple and yet cover most of the things we regard as belonging to the concept and few we regard as not belonging. The psychological literature offers a number of useful similar efforts but the best definition of intelligence we have found is the following, from the physicist van Heerden:

> Intelligent behavior is to be repeatedly successful in satisfying one's psychological needs in diverse, observably different, situations on the basis of past experience.[2]

This definition (vH) is suitable for the computer study of intelligence because it is comprehensive and its terms are not difficult to define computationally for experimental purposes. A high rate of receipt of certain reward quantities can correspond to "repeatedly successful in satisfying one's psychological needs" (on the simplest level, somatic needs). To "diverse, observably different, situations" can correspond sets of distinct sensory input "vectors" with each set having a particular implication for optimal action. To "past experience" can correspond a suitable internal record of earlier interactions with the environment, and their results.

THE ANIMAT MODEL

Computer modeling of human levels of intelligence is complex. VH's apparent applicability to both simple animals and human beings (assuming appropriate translations of its terms) suggests the usefulness of the easier course of considering basic problems that simple animals must solve, and constructing behaving models aimed at solving them. Observation of the models should aid understanding of all intelligence, and the construction of more complex models.

To define our model, we abstract four basic characteristics of simple animals:

1) The animal exists in a sea of sensory signals. At any moment only some signals are significant; the rest are irrelevant.

2) The animal is capable of actions (e.g. movement) which tend to change these signals.

3) Certain signals (e.g. those attendant on consumption of food), or certain signals' absence (e.g. absence of pain) have special status for him.

4) He acts, both externally and through internal operations, so as approximately to optimize the rate of occurrence of the special signals.

An animal's sensory-motor situation is described in very general terms by (1) and (2). Characteristics (3) and (4) are assumptions which provide a way of making definite the notion of "needs" and their satisfaction. Together, the four characteristics form the basis of our artificial animal model. For brevity, we call such a model an "animat".

We take as the animat's basic problem the generation of rules which associate sensory signals with appropriate actions so as to achieve the optimization of (4), above. For this, the major questions are adaptive, namely:

1) How to discover and emphasize rules that work,

2) Get rid of those that don't (since memory space is limited and noise is undesirable), and

3) Optimally generalize the rules that are kept (since space is limited).

There is some previous work along these lines. Notable were Grey Walter's: *machina speculatrix*, which was a sort of sub-animat which chose actions based on needs and the sensory situation, but did not adapt its rules; and *m. docilis*, which could be taught a conditioned response[3]. More recently, Holland and Reitman[4] exhibited successful performance by a rule-adaptive animat-like system which optimized its rate of satisfaction of two distinct needs. Booker[5] experimented with an animat-like "hypothetical organism" which adapted its rules in a simple environment that contained both attractive and aversive stimuli; he also provides a review of earlier systems. The present investigation is indebted to the last two works.

IMPLEMENTATION

Within the above framework we make the model definite by defining the animat's: environment, sensory channels, repertoire of actions, its association rules, and then its performance and adaptation algorithms.

Environment:

A rectangle on the computer terminal screen 18 rows by 58 columns and continued toroidally at its edges defines the environmental space. Alphanumeric characters at various positions represent objects; the animat itself is denoted by *. Some, possibly many, positions are just blank.

Sensory Channels:

In studies so far, * has been given the ability to pick up sensory signals from objects which happen to be one step (row and/or column) away, in any of the eight (including diagonal) directions; nothing is detected from more distant objects. Thus the "sense vector" has eight positions. With * located, for example, as shown below left, the sense vector would be as shown at the right:

```
T T
* F        T T F b b b b b ,
```

where b stands for blank. To form the sense vector, the circle of positions surrounding * is mapped, clockwise starting at 12 o'clock, into a left-to-right string.

But this vector is not the final sensory input. We imagine that an object is ultimately sensed as the outcome of measurements upon it by one or more feature or attribute detectors. Without loss of generality we assume each detector produces either a 0 or 1 output. If there are d detector types, an object translates into a binary string d bits in length. The sense vector as a whole thus translates into a "detector vector" of $8d$ bits. Detector translations or encodings of objects are fixed in *'s "low-level" sensory hardware. They are assigned at the beginning of an experiment. For example, in experiments discussed here, "F" (food) is encoded as "11"; "T" (tree or obstacle) as "01"; and "b" (open space) as "00". [The first bit might be thought of as the output of a "food smell?" detector; the second, of an "opacity" detector.] Thus the above sense vector translates into the detector vector:

$$01\ 01\ 11\ 00\ 00\ 00\ 00\ 00$$

The associative apparatus takes the detector vector as input.

Repertoire of Actions:

*'s actions are restricted to single-step moves in each of the eight directions. The directions are numbered 0-7 starting at 12 o'clock and proceeding clockwise; for example, a move in direction 3 would be south-easterly.

The animat may move, or attempt to move, to a position occupied by an object. The environment's response for each kind of object is predefined. In present experiments, if the move is into a position whose encoding is 00 (the blank object), there is no response (though the new sense vector will in general be different). If * steps into a space occupied by an object whose encoding has the first bit equal to 1, * is regarded as having eaten the object and receives a reward signal. If * tries to step toward an adjacent object whose encoding is 01, the step is not permitted to occur (a collision-like banging may be displayed).

The foregoing establish a semi-realistic situation in which sensory signals carry partial, but uncertain, information about the location of food, and avail-

17

able actions permit exploration and approach. Environmental predictability can be varied through the choice and arrangement of the objects. The number of object types which may be experimented with is limited only by the number of bits in the detector encoding scheme.

Association Rules:

For its association rules, the animat uses a rudimentary form of Holland's[6] "classifier" rule. The animat's rules each consist of a "taxon" and an "action". The taxon is a sort of template capable of matching a certain set of detector vectors. The action is some one of the available actions. The animat's classifier says, in effect, "if my taxon matches the current detector vector, then consider taking this action." It is a kind of hypothesis about what to do given a certain sensory situation (class of detector vectors). An example of a classifier would be:

0# 01 1# 0# 00 00 0# 0#   /   2   .

The matching rule requires that for any taxon position having a 0 or 1, the same value must occur in the detector vector; taxon positions with # (don't care) match unconditionally. Because of the #'s, which confer a kind of generality on the classifier, the above taxon, for example, will match 32 possible detector vectors, including the one discussed earlier.

It is worth making a few further observations about this classifier. First, it is a pretty good one because if food is present in direction 2 and the classifier matches the detector vector, the action recommended is to move in direction 2 and not some other direction! Second, in directions 0, 3, 6, and 7, the taxon only requires that the object be, in effect, non-food, it being irrelevant whether these directions have obstacles or are blank. Directions 1, 4, and 5 have not been so generalized. Broadly speaking, a classifier is more useful to the animat to the extent it is general (matches many detector vectors) without being so general that it makes too many errors (i.e., that in certain matching situations its recommended action is inappropriate).

Besides taxon and action, each classifier possesses a "strength", a quantity serving as the principal measure of a classifier's value to the animat. There may be other associated quantities, as well.

The animat keeps a classifier population [P] of fixed size. Usually, [P] is initialized by filling all the taxa with 0, 1, and # according to some random rule; actions are similarly filled in. As the animat's CRT "life" evolves, the classifier population changes, as will be described.

PERFORMANCE ALGORITHM

*'s basic cycle is one "step", within which events having purely to do with immediate behavior are very simple. First, the current detector vector is calculated. Second, [P] is searched for classifiers which match it; these form the "match set" [M]. Third, a classifier is selected from [M] using a probability distribution over the strengths of [M]'s classifier's; that is, the probability of selection of a particular classifier is equal to its strength divided by the sum of strengths of classifiers in [M]. Fourth, * moves according to the action of the selected classifier, or tries to. The environment's response to the move will be as described earlier.

It can be seen that *'s move choice tends to be the one having the greatest total strength among the [M] classifiers advocating it. Thus, overall, * first asks which classifiers of [P] "recognize" the current sensory situation, then from these tends to pick the move with the greatest associated strength. The subset of [M] consisting of classifiers whose action is the same as the chosen action is called the "action set" [A].

ADAPTATION ALGORITHM

The adaptation algorithm has three distinct aspects: 1) reinforcement of classifier strengths; 2) "genetic" operations on classifiers yielding new classifiers; and 3) direct creation of classifiers.

Reinforcement:

As discussed in the last section, a classifier's strength is a major determinant of its ability to influence *'s action and therefore performance. We consequently want strength to reflect the performance which tends to result when this classifier is in [A]. That would be straightforward if every step were rewarded: we could, for example, adjust the classifier's strength by an amount proportional to the reward. Classifiers which got bigger rewards would be stronger, thus more likely to be an [A], etc.

Realistically, however, it is usually the case that only some of an organism's actions receive a definite reward from the environment. Actions leading up to, or setting the stage for, a rewarded action are themselves not directly rewarded, but they must somehow be encouraged or the final payoff will not occur. Holland[7] addressed this problem in proposing a "bucket-brigade" algorithm in which, very briefly, 1) classifiers make payments out of their strengths to classifiers which were active on the preceding cycle, and 2) the same classifiers later correspondingly receive payments from the strengths of the next set of active classifiers. External reward goes to the final active set in the chain. In effect, a given amount of external reward will eventually flow all the way back through a reliable chain, reinforcing every precursor classifier.

Our basic implementation of this idea is as follows. On each step:

1) all classifiers in [A] have a fraction e of their strengths removed;

2) the total strength thus removed from [A] is distributed to the strengths of any classifiers in [A-1], defined as the action set in the previous step;

3) * then moves and if external reward is received it is distributed to the strengths of [A]; if external reward is not received, the classifiers of [A] replace those of [A-1].

Thus every [A] participates in general in two transactions, one paying out, the other receiving. We can write

$$S'_A = S_A - eS_A + p \quad ,$$

where S_A is [A]'s total strength on one step, S'_A its total on the next, and p is the total payoff received (either external reward or from the next [A]). If p is the same over time, S_A approaches a constant value given by p/e, so that under reasonably steady payoff conditions, S_A is an estimator of typical payoff. Similarly, the strength of any individual classifier is an estimator of its typical payoff.

The total payoffs to [A] and [A-1] are in the simplest case shared equally by the recipient classifiers. This has the consequence that the more classifiers are in, say, [A], the less payoff each gets.

Genetic Operations:

Consider two classifiers which match similar situations:

0# 01 1# 0# 00 00 0# 0#   /   2

and

0# 0# 11 01 00 0# 0# 0#   /   2

Each is good, but each still lacks something in generality since, for example, the matching requirements for 01 in bits 2-3 and 6-7, respectively, of each are perhaps unnecessarily restrictive. Suppose we make a new classifier by combining bits 5-9 of the first with bits 0-4 and 10-15 of the second. The result would be the slightly more general classifier:

0# 0# 1# 0# 00 0# 0# 0#   /   2   .

The above operation on two classifiers resembles a kind of crossing-over or recombination of chromosome parts in genetics. It is an operation in which two "parent" classifiers produce an offspring that is possibly an improvement over both of them. Another "genetic" operation, this time using just one parent, would first clone the parent, then mutate one or more of the clone's taxon positions. Other types of operations on classifier structure can be imagined (one will be discussed later). In each case the attempt is to use existing classifiers as the starting points for improved classifiers.

But the crossover points above were chosen quite carefully; otherwise the offspring might have been no

improvement, or even a retrogression (to a classifier more specific than either parent). We do not expect the animat to know where best to cut and mutate. How can we expect genetic operations to be of any use?

Holland[8] presents a mathematical theory showing that a population of individual symbol strings, in which each string can be assigned a numerical worth, will progressively increase in average worth as its members undergo reproduction, genetic operations on or among the offspring, and deletion of individuals to maintain constant population size. The key requirement is that an individual's probability of reproduction be proportional to its worth. Holland extended the theory to include classifier systems. In employing genetic operations, our animat constitutes an exploration and test of the theory.

The specific algorithm employed is as follows:

1) A first classifier $c1$ of [P] is selected with probability proportional to its strength;

2) If $c1$ is merely to be reproduced, a copy of it is made and added to [P]. To make room, some classifier is deleted;

3) If $c1$ is to be crossed with another classifier, a second, $c2$, is selected, also with probability proportional to strength, but from the subset of [P] of classifiers having the same action as $c1$. Two cut points are chosen as above, but at random, and an offspring $c3$ constructed out of the parts. $c3$ is added to [P] and some classifier is deleted.

Note that the parents are kept (unless one happens to suffer the deletion, but this is unlikely). The offspring, in effect, go into competition for payoff with the parents. Better (higher strength) offspring should proliferate more rapidly than their parents, driving them out; for worse offspring, the reverse should be the case.

"Create" Operations:

Occasionally, as * executes the performance algorigthm, a detector vector may occur that no classifier of [P] matches, i.e., the situation is unrecognized. The animat's response is to create a new, matching, classifier. A taxon is made by adding some #'s at random to the detector vector; an action is chosen randomly. The created classifier is added to [P] and one is deleted. The new classifier immediately matches the previously unrecognized situation and action occurs by the normal mechanism.

EXPERIMENTAL PROCEDURE

The animat model was designed with the vH-intelligence definition as a guide. In experiments with the model we are interested in finding procedures and parameter values that seem to give * greater rather than less vH-intelligence. For this two measures have been adopted. One is a performance measure: given an environment, how many steps does * take, on average, to find food objects? The other is a generality measure: does * evolve classifiers each tending to be useful in a number of distinct situations? Generality is important because it suggests that a high level of performance developed in one environment will carry over to a somewhat different environment.

The experimental procedure is to fix *'s methods and parameters, then have him do a large number of "problems" in a particular environment E. The measures of performance and generality are tracked. A "problem" always consists of starting * at a randomly selected blank position in E; then * moves until he eats some food, at which point the problem ends. The number of steps between start and food is recorded; a moving average of this quantity over the previous 50 problems is the performance measure, STPSAV.

To track generality, we calculate a histogram over the "periods" of all classifiers in [P]. The period of a classifier is a moving average of the number of steps by * between occurrences in [A] of this classifier. Thus a frequently used classifier will have a low period. [P] will then be general to the extent the histogram of periods is largest at low period. As [P] evolves we expect the histogram peak to move toward lower period, if [P]'s generality is increasing.

Figure 1. The Environment "WOODS7".

An environment used for many of the experiments is "WOODS7", shown in Fig. 1. Although WOODS7 may look easy, it actually contains a total of 92 distinct sense vectors, so *'s need to discover and generalize is substantial. To obtain performance baselines, we can start * randomly, then let him also move completely randomly until food (F) is bumped into. For WOODS7, the long-term average of the number of steps this takes is about 41

steps. We may also ask [9]: what is the best possible performance (if, say, the animat had human capabilities)? For every starting position, the number of steps to the nearest F can be found and averaged over all starting positions. The result for WOODS7 is 2.2 steps.

RESULTS AND DISCUSSION

Fig. 2 shows a performance curve for a combination of procedures and parameter settings that is among the best so far found. There is an initial rapid improvement within the first 1000 problems (untypically good during the first 100 problems, where STPSAV usually stays above 15), followed by very gradual improvement thereafter. The performance at 8000 problems, between 4 and 5 steps, is quite respectable compared with "perfect" (2.2 steps), especially since * has no information whatsoever until he is next to a nonblank object.

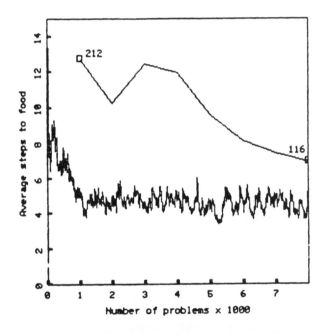

Figure 2. STPSAV (ragged line) and Period Average (broken line) for * to 8000 problems. Period values as marked.

For the same animat, Fig. 3 shows the histogram of periods of [P] at 8000 problems. There is a definite bulge for low periods; the average period is 116. For comparison, the broken line in Fig. 2 shows the trend of the period averages at earlier epochs, indicating gradual generalization in the sense we have defined.

Qualitatively, a * such as this one gives the impression of "knowing" the Woods quite well. When next to F, * nearly always takes it directly; occasionally he will move one step sideways and take it from that direction. When next to one or more T's,

20

but with no F immediately in sight, * quite reliably steps around the obstacle(s) and finds the F. When * is "out in the open", i.e., the sense vector consists of blanks, he has no information about the best way to go, as in a thick fog. One might expect *'s behavior to resemble a random walk but this is not the case. Instead, the movements look more like a general "drift" in some direction, with some superimposed randomness. After several problems the drift may shift to another direction.

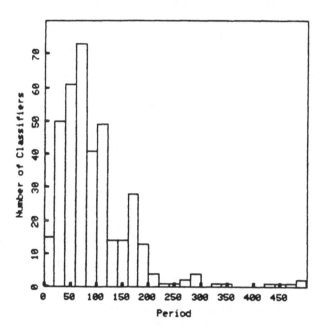

Figure 3. Histogram of classifier periods for the * of Figure 2 at 8000 problems.

Parameter Values:

Parameter values for the animat of Fig. 2 were arrived at by experiment. Three basic parameters are discussed in this section, with observations about setting them reasonably.

For Fig. 2, [P] contained 400 classifiers. A suitable value for this number appears related to the number of distinct sense vectors or "scenes" (here, 92) in the environment. Too small a ratio of classifiers to scenes results in "forgetful" behavior in which * keeps losing good moves that appeared well learned. A small ratio means that for some scenes deletion has a high probability of eliminating all matching classifiers. For ratios above about four, the forgetting is much less noticeable. To the extent * generalizes, more and more classifiers match each sense vector, further reducing the problem.

The "estimator fraction", e, was set at 0.2, i.e., a classifier lost 20 percent of its strength each time it entered [A]. In general, smaller values of e mean that a classifier's strength reflects a weighted average of payoffs that reaches farther into the past. Conversely, a larger value makes the strength more sensitive to recent payoffs. It was found that $e = 0.4$ produced a noticeably more erratic STPSAV curve, whereas changing from $e = 0.2$ to 0.1 did not affect the curve significantly. Strength should accurately estimate a classifier's typical payoff. In this problem, payoff fluctuations are apparently large enough so that $e = 0.4$ results in too short an averaging interval for good estimation. If e is too small, though, newly formed classifiers may get evaluated too slowly; we therefore kept e at 0.2.

The rate at which genetic operations occurred was set proportional to the problem rate. Specifically, at the end of each problem, a single genetic event (as described earlier) took place with probability RGPROB. Given the event, crossover occurred with probability XPROB. Settings were typically 0.25 and 0.50, respectively. These seemed to ensure that, on average, classifiers would be fully evaluated by the reinforcement process by the time they were selected for a genetic operation (or deleted). Typically, a problem took five steps in which each set [A] had about 10 members, giving about 50 evaluations. The above value for RGPROB then implies 200 evaluations per genetic event. This seems excessive except that some classifiers are much more frequently used than others and we wanted to allow for the well-rewarded but infrequently called-upon classifier. It is possible our results would have been speeded up, without adverse side effects, by a higher genetic rate.

Distance Estimation:

Performance in the earliest animat experiments was far below the level of Fig. 2. One defect was a kind of "dithering" in which while * would tend toward F's, the path would have unnecessary sidesteps and wanderings. It was then realized that the basic reinforcement algorithm does not care whether a path from point A to food is long or short; there is nothing which preferentially reinforces the most expeditious classifiers. Any path, even a looping one, will come to equilibrium at a high strength level in its constituent classifiers.

The solution had to be more subtle than simply penalizing long paths. What is required is a technique that, at every position, tends to prefer the most direct of several possible moves, but does not prevent the setting up of a long path if that is actually the shortest path available. Our solution was twofold. First, each classifier was made to keep an estimate of its distance (in steps) to food. This did not require elaborate look-ahead. Instead each classifier in [A-1] adjusted its distance estimate according to an average of the distance estimates of [A]; when reward was received, the members of [A] were similarly adjusted, using the quantity 1. This tech-

21

nique, with each estimate an average over the last few updates, is quite satisfactory.

The distances are employed as follows. In the performance cycle, selection from [M] is based on probability proportional to strength/distance instead of just strength. Consequently, a move tends to be selected that is not only strong, but also "short". Now comes the second part of the solution. At the same time as [A] is formed, the set NOT[A] of the remaining classifiers in [M] is taxed by a small amount (typically five percent): the "longer" classifiers thus tend to incur a loss by not being selected. This "lateral inhibition" induces a sort of catastrophe in which the shorter classifiers become even more likely to be picked and the longer become ever weaker, and can disappear entirely. Note that the competition is purely local and does not work against the setting up of minimal long paths.

This technique is very effective against "dithering"; the progressive takeover of a match set by a discovered shorter move has been repeatedly observed. Our solution is not perfect, however, because to suppress the special case of occasional looping situations we had to impose a small tax (five percent) on [A]. Since [A] is the set which receives payoff, the tax has little effect except if a loop is taking place, and then the tax is soon very effective. Still, in principal, even a small tax on [A] reduces the strength flow in very long chains, putting them at a reproductive disadvantage. This residual problem may be an indication that as paths grow, they should be "condensed" into units of behavior longer than one step.

Extensions to "Create"

A second area of changes which improved performance had to do with the "Create" operations. As discussed, Create at first only occured when [M] was empty. It was found that * sometimes also got stuck looping among situations with nonempty [M]'s. The tax on [A] enabled recognition of these loops because the total strengths in each [A] would tend to zero. We put in a threshold that triggered Create if the strength of any [M] got too low. This suppressed looping dramatically and improved performance.

It was also found important to trigger Create randomly, at a very low rate (typically, with probability 0.02 per step). * is engaged in path construction, using the best available current evidence. This can lead to good but nevertheless suboptimal paths which might be improved if * would only try something different. Random Creates are one way to introduce a new move direction. Usually the new classifier is no improvement. But when it is, and it gets tried (gets in [A]), it will be (often heavily) reinforced and therefore given a good chance at eventual reproductive success.

A different type of Create was also found useful. Instead of randomly picking the action in a Created classifier, * may make an educated guess, as follows. From its current position, * steps tentatively into a randomly selected adjacent position. There, [M] is determined and the strength-weighted average of the distances of its classifiers, MNDIST[M], is formed. The same is done for several adjacent positions. These values are then compared with MNDIST[M] for the starting position. Several decision schemes are possible, with the general idea of picking an action direction corresponding to the shortest apparent path. If, however, none of the adjacent MNDIST[M]'s is better by more than 1 than the current position's value, it is preferable not to create a new classifier. This technique is important early in *'s existence, when very little is yet known; but, interestingly, it appears that * should not rely entirely upon it. Some suboptimal paths get set up which tend not to be improved. The problem goes away if random Creates are also available.

Effect of Genetic Operations:

Finally, we shall discuss what the experiments suggest about the role of the genetic operations. To begin, it is helpful to define a "concept" as a set of classifiers from [P] having exactly the same taxon and action, and for which there is no other classifier in [P] with that taxon and action. The basic effect of *'s genetic operations then appears to be to exert a pressure tending to increase the generality of [P]'s concepts. That is, with time, the periods of the concepts in [P] tend to decrease. The pressure is restrained by the requirement that the concepts be more or less correct (* must get the food expeditiously). The precise point of balance appears to depend on the parameter regime.

An important experiment is to evolve an animat with reinforcement and Create going as usual, but with genetic operations turned off. The result is a performance almost as good as Fig. 2. But significant generalization does not occur; the curve of histogram averages remains essentially flat at a value of about 270. There thus appears to be a division of effort: Create introduces the raw material, the specific examples to be evaluated; and the genetic operations produce more general concepts from the examples.

It is clear that crossover is capable of making a more general classifier out of two less general parents; this was illustrated earlier. We are not sure, however, just why for * the more general concept has a selective advantage. Somehow, greater generality must lead to greater concept strength; there is no other way to win out. Yet being active more frequently does not in itself result in greater strength: strength is an estimator typical payoff, not payoff rate.

Our tentative hypothesis stems from noting that a more specific concept will always have to share payoff with any more general offspring that comes into existence. This initially weakens the specific concept so that the number of classifiers making it up tends to fall (at equilibrium, numbers are proportional to total strength). Consequently, the specific gets even less of the payoff, since payoff is shared. The result is a cascading situation in which the more general concept wins out. The odds favor the general because it has more than this one source of payoff.

While general classifiers appear to have a selective advantage, this is of no use unless such classifiers can be formed and introduced in the first place. Crossover is adequate for some types of generalization. But a natural operation for the purpose is obviously intersection. We have implemented this operation as follows. Two parents are chosen and a new taxon is formed by intersecting copies of the parents' taxa over a randomly selected interval. In that interval, if the parents differ at a position, the new taxon gets a #; if not, the new taxon gets the common value. Outside the interval, the new taxon is filled in from parent 1.

Intersection is a "hot" operation which should be used cautiously because it can introduce #'s at a high rate. Nevertheless, our results show increased generalization with little performance loss when crossover and intersection are both available to *.

Space remains only discuss the deletion technique. The simplest method, conceptually, is to delete at random. Then, to a first approximation, the equilibrium number of classifiers in a concept—or in any subset of [P] whatsoever—is proportional to its total strength. A drawback of random deletion is that a valuable concept that happens to consist of one classifier is at considerable risk until it reproduces. This is not a problem on average if [P] is large enough. Yet one wonders whether "deleting the weak" might not be better.

Several methods have been tried, all but one clearly worse than random deletion. The possibly better method is to delete with probability proportional to the reciprocal of strength. This has the obvious effect of tending to protect the precious classifier just mentioned. It can also be shown that the probability that a concept [C] will lose a member under this type of deletion is proportional to the square of its number, which places a strong restraint on over-expansion.

The * of Fig. 2 employed both intersection (along with crossover) and inverse-strength deletion.

CONCLUSION

In its simple way, * meets the definition of intelligence stated at the beginning. * becomes good at satisfying its need for food in a Woods of diverse object configurations on the basis of experience. Though not yet tested, *'s rule generalization over time suggests that performance would be maintained in a somewhat different Woods, or if the Woods slowly changed.

While the present animat has numerous limitations (sensory, motor, memory, etc.) there does not seem to be any essential barrier to removal of the limitations and to carryover of the present algorithms to a more sophisticated model in more complicated environments.

ACKNOWLEDGEMENT

The author wishes to acknowledge valuable conversations with C.G. Shaefer of the Rowland Institute.

REFERENCES

[1] Winston, P.H. *Artificial Intelligence*, 2nd ed. Reading, Massachusetts: Addison-Wesley, 1984.

[2] van Heerden, P.J. *The Foundation of Empirical Knowledge.* Wassenaar, The Netherlands: Wistik, 1968.

[3] Walter, W.G. *The Living Brain.* New York: Norton, 1953.

[4] Holland, J.H., & Reitman, J.S. Cognitive systems based on adaptive algorithms. In *Pattern-Directed Inference Systems*, Waterman, D.A., & Hayes-Roth, F., (eds.). New York: Academic Press, 1978.

[5] Booker, L. *Intelligent Behavior as an Adaptation to the Task Environment*, Ph.D. Dissertation (Computer and Communication Sciences), The University of Michigan, 1982.

[6] Holland, J.H. Adaptation. In *Progress in Theoretical Biology, 4*, Rosen, R., & Snell, F.M., (eds.). New York: Plenum, 1976.

[7] ——— . Genetic algorithms and adaptation. In *Adaptive Control of Ill-Defined Systems*, Selfridge, O.G., Rissland, E.L., & Arbib, M.A., (eds.). New York: Plenum, 1984.

[8] ——— . *Adaptation in Natural and Artificial Systems.* Ann Arbor: University of Michigan Press, 1975.

[9] Martha Gordon, personal communication.

IMPLEMENTING SEMANTIC NETWORK STRUCTURES

USING

THE CLASSIFIER SYSTEM

Stephanie Forrest
The University of Michigan
Ann Arbor, Michigan

Introduction

One common criticism of Classifier Systems is the low-level nature of their representations. In Classifier Systems information is stored as rules (classifiers) that have a very constrained format (binary bit strings). Low-level binary bit string representations support adaptive learning algorithms well (Holland, 75)(Holland, 80). However, it is difficult to interpret the behavior of these systems without a high-level interpreter that can code and de-code the ones and zeroes into more meaningful terms. In particular, although gross behaviors can be measured at various intervals using some fitness function it is difficult to chart how learning takes place or to determine what role is played by each component of the system. This feature of low-level representations makes it difficult to establish direct connections between the behavior of Classifier Systems and more common high-level symbolic representations used in artificial intelligence programs.

The research described in this paper addresses this criticism by demonstrating that Classifier Systems are capable of representing sophisticated high-level structures. This has been accomplished by selecting one class of knowledge representation paradigms (semantic networks) and showing how they can be implemented as a collection of Classifier System rules. The described system takes high-level semantic network descriptions as input and automatically translates them into a Classifier System representation. It also provides a "query processor" that takes high-level queries about the semantic network, translates them into a sequence of Classifier System operations, and translates the results of the queries back

24

into higher-level answers.

In large scale parallel systems such as the Classifier System, the issue of control is central. Control issues arise in two ways for the Classifier System: in deciding which external classifiers are to be generated, and in deciding which external messages are to be placed on the message list and when. As the number of rules in the system increases, it quickly becomes impossible to do control the system manually. There are at least two possible ways to automate the process: "learning" and "compiling." Compilation can be viewed as mapping high-level structures onto lower-level operations ("top down"). Likewise, some kinds of learning (for example, genetic algorithms) can be viewed as the gradual emergence of higher-level structures from a random assortment of low-level processes; systems using these kinds of learning organize themselves from the "bottom up." The bottom-up approach is the one that has been studied previously for Classifier Systems (Holland, 80) (Booker, 82) (Goldberg, 83). The top-down approach is being explored in this paper.

The implementation takes the form of a compiler, mapping "high-level" semantic network definitions onto the Classifier System. In this context, the Classifier System is properly viewed either as a lower-level target language or as a specification for an abstract parallel machine. One particular semantic network formalism was selected for this research: KL-ONE (Brachman, 78) (Schmolze and Brachman, 82) (Brachman and Schmolze, 85). The KL-ONE family of languages is widely used; it contains most of the common semantic network constructs (the most notable exception being cancel links), has been precisely described, and includes sophisticated accessing functions as part of the design of the language. These characteristics make KL-ONE an excellent exemplar of the semantic network representation paradigm.

The remainder of this paper is divided into five sections: (1) brief description of my version of the Classifier System, (2) short introduction to KL-ONE, (3) description of the Classifier System implementation of KL-ONE, (4) discussion, and (5) conclusions.

25

The Classifier System

Since there are several variants of Classifier Systems, I will describe below the one used for this project. This particular system does not include those features that are specific to the use of adaptive algorithms, such as bidding, support, etc. This is because I am interested in showing what sorts of representations are possible, not how they can evolve. The following view of the Classifier System emphasizes how it can be used to represent higher-level structures and does not rely on any particular hardware implementation. Thus, it is appropriate to describe the language of possible programs for the Classifier System as a formal grammar. The input to a Classifier program is the set of external messages (often called detector messages) that are added to the message list during the program's execution. The output is the set of messages (called effector messages) read from the message list by an external agent. Just as many traditional programs can be run interactively, a classifier program can be thought of as receiving intermittent input from the external environment and occasionally emitting output messages. The syntax for the Classifier System is as follows:

$$<\text{classifier system}> ::= <\text{classifier}>^{*}$$

$$<\text{classifier}> ::= <\text{condition}>^{+} => <\text{action}>$$

$$<\text{condition}> ::= <\text{alphabet}>^{n} \mid \sim<\text{alphabet}>^{n}$$

$$<\text{action}> ::= <\text{alphabet}>^{n}$$

$$<\text{alphabet}> ::= 1 \mid 0 \mid \#$$

Each classifier, or production rule, consists of a condition part and an action part. The action part specifies exactly one action, while the condition part may contain many conditions (pre-conditions of activation). Rules with more than one condition are referred to as "multiple-condition classifiers." A multiple-condition classifier must have each of its pre-conditions fulfilled in a single time step for it to be activated. The conditions and actions are fixed length strings over the alphabet $(1,0,\#)$ where $\#$ denotes "don't care" and 1 and 0 are literals. The determination of whether or not a specific message matches a condition is a

26

logical bit comparison on the defined (1 or 0) bits. If a "not" condition is used, the condition is fulfilled just in the case that no message on the message list matches it. The #'s in the condition part designate "don't care" positions in the sense that they match either 1 or 0. The action part of the classifier determines the message to be posted. All defined bits appear directly in the output message. Any # symbols in the action part indicate that the corresponding bit value in the activating message should be substituted for the # symbol in the output message.[1] Actual messages are always completely defined in that they do not contain "don't care" symbols. Separate conditions are placed on separate lines, and the first condition (the distinguished condition) of a classifier is used to pass through messages to the action part.

As a simple example, consider the following four bit (n = 4) classifier system:

$$\#00\# => 1101$$

$$\#101$$
$$\#\#\#1 => \#\#1\#$$

$$\sim 1111 => 1111.$$

This classifier system has three classifiers. The second classifier illustrates multiple-conditions, and the third contains a negative condition. If an initial message, "0000" is placed on the message list at time T0, the pattern of activity shown below will be observed on the message list:

Time Step	Message List	Activating Classifier
T0:	0000	external
T1:	1101	first
	1111	third
T2:	1111	second
T3:		
T4:	1111	third.

---

[1]For multiple-condition classifiers, this operation is ambiguous since it is not clear what it means to simultaneously perform "pass through" on more than one condition. The ambiguity is resolved by selecting one condition to be used for pass through. By convention, this will always be the first condition. Another ambiguity arises if more than one message matches the distinguished condition in one time step. Again by convention, in my system I process all the messages that match this condition. The example illustrates this procedure.

The final two message lists (null and "1111") would continue alternating until the system was turned off. In T1, one message (1101) matches the first (distinguished) condition and both messages match the second condition. Pass through is performed on the first condition, producing one output message for time T2. If the conditions had been reversed (###1 distinguished), the message list at time T2 would have contained two identical messages (1111).

KL-ONE

KL-ONE organizes descriptive terms into a multi-level structure that allows properties of a general concept, such as "mammal," to be inherited by more specific concepts, such as "zebra." This allows the system to store properties that pertain to all mammals (such as "warm-blooded") in one place but to have the capability of associating those properties with all concepts that are more specific than mammal (such as zebra). A multi-level structure such as KL-ONE is easily represented as a graph where the nodes of the graph correspond to concepts and the links correspond to relations between concepts. Such graphs, with or without property inheritance, are often referred to as semantic networks.

KL-ONE resembles NETL [Fahlman, 79] and other systems with default hierarchies in its exploitation of the idea of structured inheritance of properties. It differs by taking the definitional component of the network much more seriously than these other systems. In KL-ONE, properties associated with a concept in the network are what constitute its definition. This is a stronger notion than the one that views properties as predicates of a "typical" element, any one of which may be cancelled for an "atypical" case. KL-ONE does not allow cancellation of properties. Rather, the space of definitions is seen as an infinite lattice of all possible definitions: there are concepts to cover each "atypical" case. All concepts in a KL-ONE network are partially ordered by the "SUBSUMES" relation. This relation, often referred to as "IS-A" in other systems, defines how properties are inherited through the network. That is, if a concept A is subsumed by another concept B, A inherits

28

all of B's properties. Included in the lattice of all possible definitions are contradictory concepts that can never have an extension (instance) in any useful domain, such as "a person with two legs and four legs." Out of this potentially infinite lattice, any particular KL-ONE network will choose to name a finite number of points (because they are of interest in that application), always including the top element, often referred to as "THING."

KL-ONE also provides a mechanism for using concepts whose definitions either cannot be completely articulated or for which it is inconvenient to elaborate a complete definition — the PRIMITIVE construct. For example, if one were representing abstract data types and the operations that can be performed on them, it might be necessary to mention the concept of "Addition." However, it would be extremely tedious and not very helpful in this context to be required to give the complete set-theoretic definition of addition. In a case such as this, it would be useful to define addition as a primitive concept. The PRIMITIVE construct allows a concept to be defined as having something special about it beyond its explicit properties. Concepts defined using the PRIMITIVE construct are often indicated with "*" when a KL-ONE network is represented as a graph.

While NETL stores assertional information (e.g., "Clyde is a particular elephant") in the same knowledge structure as that containing definitional information (for example, "typical elephant"), KL-ONE separates these two kinds of knowledge. A sharp distinction is drawn between the definitional component, where terms are represented, and the assertional component, where extensions (instances) described by these terms are represented. It is possible to make more than one assertion about the same object in any world. For example, it may be possible to assert that a certain object is both a "Building" and a "Fire Hazard." In KL-ONE, the definitional component (and its attendant reasoning processes) of the system is called the "terminological" space, and a collection of instances (and the reasoning processes that operate on it) is referred to as the "assertional" space. The features of KL-ONE that are discussed here (structured inheritance, no cancellation of properties, primitive concepts, etc.) reside in the terminological component, while statements in the assertional component are represented as sentences in some defined logic. Reasoning in the assertional part of the system is generally viewed as theorem proving.

29

At the heart of knowledge acquisition and retrieval is the problem of classification. Given a new piece of information, classification is the process of deciding where to locate that information in an existing network and knowing how to retrieve it once it has been entered. This information may be a single node (concept) or, more likely, it may be a complex description built out of other concepts. Because KL-ONE maintains a strict notion of definition, it is possible to formulate precise rules about where any new description (terminological) should be located in an existing knowledge base.

As an example of this classification process in KL-ONE, if one wants to elaborate a new concept XXXX that has the following characteristics:

(1) XXXX is a kind of vacation,
(2) XXXX takes place in Africa, and
(3) XXXX involves hunting zebras,

there exists a precise way to determine which point in the lattice of possible definitions should be elaborated as XXXX.[2] Finding the proper location for XXXX would involve finding all subsumption relationships between XXXX and terms that share characteristics with it.

If the terminological space is implemented as a multi-level network, this process can be described as that of finding those nodes that should be immediately above and immediately below XXXX in the network. The notions of "above" and "below" are expressed more precisely by the relation "SUBSUMES." Deciding whether one concept SUBSUMES another is the central issue of classification in KL-ONE. The subsumption rules for a particular language are a property of the language definition (Schmolze and Israel, 83).

In summary, there are two aspects to the KL-ONE system: (1) data structures that store information and (2) a sophisticated set of operations that control interactions with those data structures. In the following sections, the first of these aspects is emphasized. A more detailed treatment of KL-ONE operations is contained in (Lipkis, 81).

---

[2]More precisely, XXXX has a location role which is value restricted to the concept Africa, an activity role which is value restricted to concept HuntingZebras, and a SUPERC link connecting it to the concept Vacation.

Classifier System Implementation of KL-ONE

In this section, a small subset of the KL-ONE language is introduced and the corresponding representation in classifiers is presented. Then it is shown how simple queries can be made to the Classifier System representation to retrieve information about the semantic network representation. The simple queries that are discussed can be combined to form more complex interactions with the network structure (Forrest, 85).

A KL-ONE semantic network can be viewed as a directed graph that contains a finite number of link and node types. Under this view, a Classifier System representation of the graph can be built up using one classifier to represent every directed link in the graph. The condition part of the classifier contains the encoded name of the node that the link comes from and the action part contains the encoded name of the node that the link goes to. Tagging controls which type of link is traversed. In the following, two node types (concepts and roles) and six link types (SUPERC, ROLE, VR, DIFF, MAX, and MIN) are discussed. These node and link types comprise the central core of most KL-ONE systems and are sufficiently rich for the purposes of this paper.

For the purposes of encoding, the individual bits of the classifiers have been conceptually grouped into fields. The complete description of these fields appears below. The description of the encoding of KL-ONE is then presented in terms of fields and field values, rather than using bit values. It should be remembered that each field value has a corresponding bit pattern and that ultimately each condition and action is represented as a string of length thirty-two over the alphabet $\{1,0,\#\}$. The word nil denotes "don't care" for an entire field. There are several distinct ways in which the classifiers' bits have been interpreted. The use of tagging ensures that there is no ambiguity in the interpretations used. The type definition facilities of Pascal-like languages provide a natural way to express the the conceptual interpretations I have used, as shown below:

```
type
  tag = (NET,NUM,PRE);
  link = (SUPERC,ROLE,DIFF,VRLINK,MAX,MIN,);
  direction = (UP,DOWN);
  compare = (AFIELD,BFIELD,CFIELD);
  name = string;
  message = string;
  numeric = 0 .. 63;

  classifier pattern = record
  case tag : tagfield

      NET       : /* Structural Variant */
            (tagfield name link direction);

      NUM       : /* Numeric Variant */
            (tagfield name nil direction compare numeric);

      PRE       : /* PreDefined Variant */
            (tagfield message);
  end;
```

This definition defines three patterns for constructing classifiers: structural, numeric, and predefined. The structural pattern is by far the most important. It is used to represent concepts and roles. The numeric pattern is used for processing number restrictions. The predefined pattern is used for control purposes; it has no don't cares in it, providing reserved words, or constants, to the system.

The structural pattern has been broken into four fields: tag, name, link, and direction. The tag field is set to NET, the name field contains the coded name of a concept or role, the link field specifies which link type is being traversed (SUPERC, DIFF, etc.), and the direction determines whether the traversal is up (specific to general) or down (general to specific).

The Numeric pattern has six fields: tag, name, link, direction, compare, and number. In most cases the name, link, and direction fields are not relevant to the numeric processing and are filled with don't cares. The tag field is always set to NUM, and the compare field is one of AFIELD, BFIELD, or CFIELD. The compare field is used to distinguish operands in arithmetic operations. The number field contains the binary representation of the number being processed.

The Predefined pattern has the value PRE in the tag field. The rest of the pattern is assigned to one field. These bits are always completely defined (even in conditions and

32

actions) as they refer to unique constant messages. These messages provide internal control information and they are they are used to initiate queries from the command processor.

Concept Specialization

All concepts in KL-ONE are partially ordered by the "SUBSUMES" relation. One concept, for example Surfing, is said to specialize another concept, say WaterSport, if Surfing is SUBSUMEd by WaterSport. This means that Surfing inherits all of WaterSport's properties. The "SUBSUMES" relation can be inferred by inspecting the respective properties of the two concepts, or Surfing can be explicitly defined as a specialization of WaterSport. Graphically, the specialization is represented by a double arrow (called a SUPERC link) from the subsumed concept to the subsuming concept (see Figure 1). KL-ONE's SUPERC link is often called an ISA link in other semantic network formalisms. Since the SUBSUMES relation is transitive, SUPERC links could be drawn to all of WaterSport's subsumers as well. Traditionally, only the local links are represented explicitly.

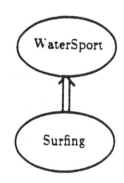

Figure 1
Concept Specialization

Two classifiers are are needed to represent every explicit specialization in the network. This allows traversals through the network in either the UP (specific to general) or DOWN (general to specific) direction. The classifiers form the link between the concept that is being specialized and the specializing concept. The following two classifiers represent the

33

network shown in Figure 1:

NORM–WaterSport–SUPERC–DOWN => NORM–Surfing–SUPERC–DOWN

NORM–Surfing–SUPERC–UP => NORM–WaterSport–SUPERC–UP.

Roles

A role defines an ordered relation between two concepts. Roles in KL-ONE are similar to slots in frame-based representations. The domain of a role is analogous to the frame that contains the slot; the range of a role is analogous to the class of allowable slot-fillers. In KL-ONE, the domain and range of a role are always concepts. Just as there is a partial ordering of concepts in KL-ONE, so is there a partial ordering of roles. The relation that determines this ordering is "differentiation." Pictorially, the DIFFERENTIATES relation between two roles is drawn as a single arrow (called a DIFF link). Roles are indicated by a circle surrounding a square (see Figure 2). This allows roles to be defined in terms of other roles similarly to the way that concepts are defined from other concepts. The domain of a role is taken to be the most general concept at which it is defined, and, likewise, the range is taken to be the most general concept to which the role is restricted (called a value restriction). If there is no explicit value restriction in the network for some role, its range is assumed to be the top element, THING.

Roles are associated with a concept, and one classifier is needed to represent each association (link) between a concept and its role. For example, the role Arm might be associated with the concept Person (see Figure 2) and the following classifier would be generated:

nil–Person–nil–nil–nil
PRE–RoleMessage => nil–Arm–DIFF–nil–nil.

Roles can be defined in terms of other roles using DIFF links. For example, the role Sibling can be defined as a differentiater of "Relatives" (see Figure 3). Building on this definition, the conjunction WealthySibling is defined by constructing DIFF links from WealthySibling both to Sibling and to Wealthy as shown in Figure 3.

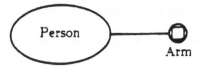

Figure 2
Concept and Role

Figure 3 shows how these would be drawn.

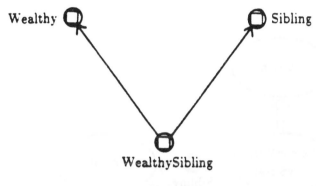

WealthySibling

Figure 3
Role Differentiation

There are two links specified by this definition. Two classifiers are needed to represent each link so that queries can be supported in both directions (UP or DOWN). They are shown below:

    NORM-Wealthy-DIFF-DOWN => NORM-WealthySibling-DIFF-DOWN

    NORM-WealthySibling-DIFF-UP => NORM-Wealthy-DIFF-UP

    NORM-Sibling-DIFF-DOWN => NORM-WealthySibling-DIFF-DOWN

    NORM-WealthySibling-DIFF-UP => NORM-Sibling-DIFF-UP.

These classifiers control propagations along DIFF links. They could be used to query the system about relations between roles.

Value Restrictions

Value restrictions limit the range of a role in the context of a particular concept. In

35

frame/slot notation, this would correspond to constraining the class of allowable slot fillers for a particular slot. To return to the sibling example, we might wish to define the concept of a person all of whose siblings are sisters (PersonWithOnlySisters). In this case the role, Sibling, is a defining property of PersonWithOnlySisters. The association between a concept and a role is indicated in the graph by a line segment connecting the concept with the role. Value restrictions are indicated with a single arrow from the role to the value restriction (a concept). Figure 4 illustrates these conventions.

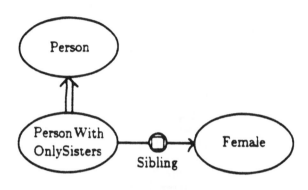

Figure 4
Value Restrictions

One classifier is needed for each explicitly mentioned value restriction. This classifier associates the local concept and the relevant role with their value restriction. The control message, VR, ensures that the classifier is only activated when the system is looking for value restrictions. The following classifier is produced for the value restriction:

```
nil-PersonWithOnlySisters-nil-nil-nil
nil-Sibling-nil-nil-nil
PRE-VRMessage => nil-Female-SUPERC-nil-nil.
```

It should be noted that the above definition does not require a PersonWithOnlySisters to actually have any siblings. It just says that if there are any, they must be female. The definition can be completed to require this person to have at least one sister by placing a number restriction on the role.

Pictorially, number restrictions are indicated at the role with (x,y), where x is the lower bound and y is the upper bound. Not surprisingly, these constructs place limitations on the minimum and maximum number of role fillers that an instance of the defined concept can have. In KL-ONE, number restrictions are limited to the natural numbers. The default MIN restriction for a concept is zero, and the default MAX restriction is infinity. Thus, in the above example, the concept PersonWithOnlySisters has no upper bound on the number of siblings.

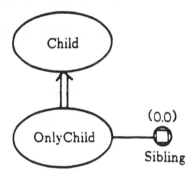

Figure 5
Number Restrictions

Consider the definition of an only child, shown in Figure 5. This expresses the definition of OnlyChild as any child with no siblings. The following two classifiers would be generated for the number restriction:

```
nil-Sibling-nil-nil-nil
nil-OnlyChild-nil-nil-nil
PRE-MaxMessage => NUM-nil-MAX-nil-nil-0

nil-Sibling-nil-nil-nil
nil-OnlyChild-nil-nil-nil
PRE-MinMessage => NUM-nil-MIN-nil-nil-0.
```

Querying The System

Four important KL-ONE constructs and their corresponding representations in

37

classifiers have been described. These are: concept specialization, role attachment and differentiation, value restriction, and number restriction. Once a Classifier System representation for such a system has been proposed, it is necessary to show how such a representation could perform useful computations. In particular, it will be shown how the collection of classifiers that represent some network (as described above) can be queried to retrieve information about the network. An example of such a retrieval would be discovering all the inherited roles for some concept.

In the context of the Classifier System, the only IO capability is through the global message list. The form of a query will therefore be a message(s) added to the message list from some external source (a query processor) and the reply will likewise be some collection of messages that can be read from the message list after the Classifier System has iterated for some number of time steps.

As an example, consider the network shown in Figure 6 and suppose that one wanted to find all the inherited roles for the concept HighRiskDriver. First, one new classifier must be added to the rule set:

NET–nil
~ PRE–ClearMessage => NET–nil.

This classifier allows network messages to stay on the message list until it is explicitly deactivated by a ClearMessage appearing on the message list.

The query would be performed in two stages. First, a message would be added to the message list that would find all the concepts that HighRiskDriver specializes (to locate all the concepts from which HighRiskDriver can inherit roles). This query takes two time steps. After the second time step (when the three concepts that HighRiskDriver specializes are on the message list), the second stage is initiated by adding the "Role" message to the message list. It is necessary at this point to ensure that the three current messages will not be rewritten at the next time step so that the role messages will not be confused with the concept messages. This is accomplished by adding the ClearMessage, which "turns off" the one overhead classifier. Both stages of the query are shown below:[3]

---

[3]The -> symbol indicates messages that are written to the message list from an

38

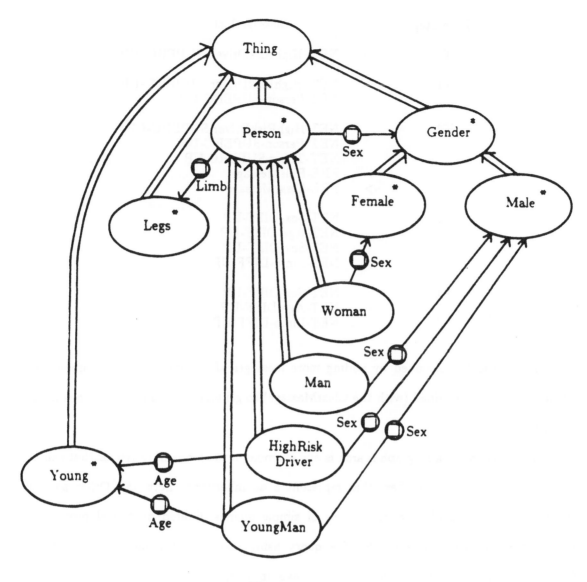

Figure 6
Example KL-ONE Network.

external source.

39

Time Step		Message List
T0:	->	NET–HighRiskDriver-SUPERC–UP
T1:		NET–HighRiskDriver-SUPERC–UP
		NET-Person-SUPERC-UP
T2:		NET-HighRiskDriver-SUPERC-UP
		NET-Person-SUPERC-UP
		NET-Thing-SUPERC-UP
	->	PRE-RoleMessage
	->	PRE-ClearMessage
T3:		NET-Sex-DIFF-UP
		NET-Age-DIFF-UP
		NET-Sex-DIFF-UP
		NET-Limb-DIFF-UP
T4:		NET-Sex-DIFF-UP
		NET-Age-DIFF-UP
		NET-Limb-DIFF-UP.

The query could be continued by adding more messages after time T4. For example, the VRMessage could be added (with the ClearMessage) to generate the value restrictions for all the roles on the list.

This style of parallel graph search is one example of the kinds of retrievals that can be performed on a set of classifiers that represent a an inheritance network. Other parallel operations include: boolean combinations of simple queries. limited numerical processing, and synchronization. An example of a query using boolean combinations would be to discover all the roles that two concepts have in common. This is accomplished by determining the inherited roles for each of the two concepts and then taking their intersection. Queries about number restrictions involve some numerical processing. Finally, it is also possible to synchronize the progression of independent queries. For these three types of queries, additional overhead classifiers are required.

Discussion

The techniques discussed in the previous section have been implemented and fully described (Forrest, 85). These techniques are presented in the context of more complex KL-

ONE operations such as classification and determination of subsumption.

The implemented system (excluding the Classifier System simulation) is divided into four major parts: parser, classifier generator, symbol table manager, and external command processor. The parser takes KL-ONE definitions as input, checks their syntax, and enters all new terms (concepts or roles) into a symbol table. The classifier generator takes syntactically correct KL-ONE definitions as input and (using the symbol table) constructs the corresponding classifier representation of the KL-ONE expression. The parser and classifier generator together may be thought of as a two pass compiler that takes as input KL-ONE network definitions and produces "code" (a set of classifiers) for the Classifier System. Additional classifiers that are independent of any given KL-ONE network (for example, the overhead classifier described in the previous section) are loaded into the list of network classifiers automatically. These include classifiers to perform boolean set operations, sorting, arithmetic operations, etc. The symbol table contains the specific bit patterns used to represent each term in a KL-ONE definition. One symbol table is needed for each KL-ONE network. Thus, if new concepts are to be added to a network without recompilation, the symbol table must be preserved after "compilation." The external command processor runs the Classifier System, providing input (and reading output) from the "classifier program."

Several techniques for controlling the behavior of a Classifier System have been incorporated into the implementation. Tagging, in which one field of the classifier is used as a selector, is used to maintain groups of messages on the message list that are in distinct states. This allows the use of specific operators that are defined for particular states. This specificity also allows additional layers of parallelism to be added by processing more than one operation simultaneously. In these situations, the messages for each operation are kept distinct on the global message list by the unique values of their tags.

Negative conditions activate and deactivate various subsystems of the Classifier System. Negative conditions are used to terminate computations and to explicitly change the state of a group of messages when a "trigger" message is added to the list. The trigger condition violates the negative condition and that classifier is effectively turned off.

41

Computations that proceed one bit at a time illustrate two techniques: (1) using control messages to sequence the processing of a computation, and (2) how to collect and combine information from independent messages into one message. Sequencing will always be useful when a computation is spread out over multiple time steps instead of being performed in one step. Collection is important because in the Classifier System it is easy to "parallelize" information from one message into many messages that can be operated on independently. This is most easily accomplished by having many classifiers that match the same message and operate on various fields within the message. The division of one message into its components takes one time step. However, the recombination of the new components back into one message (for example, an answer) is more difficult. The collection process must either be conducted in a pairwise fashion or a huge number of classifiers must be employed. The computational tradeoff for n bits is 2^n classifiers (one for each combination of possible messages) in one time step versus n classifiers (one for each bit) that are sequenced for n time steps. Intermediate solutions are also possible.

Synchronization techniques allow one operation to be delayed until another operation has reached some specific stage. Then both operations can proceed independently until the next synchronization point. Synchronization can be achieved by combining tagging with negative conditions.

Conclusions

Classifier Systems are capable of representing complex high-level knowledge structures. This has been shown by choosing one example of a common knowledge representation paradigm (KL-ONE) and showing how it can be translated into a Classifier System rule set. In the translation process the Classifier System is viewed as a low-level target language into which KL-ONE constructs are mapped. The translation is described as compilation from high-level KL-ONE constructs into low-level classifiers.

Since this study has not incorporated the bucket brigade learning algorithm, one obvious direction for future study is exploration of how many of the structures described here are learnable by the bucket brigade. This would test the efficacy of the learning algorithm and it would allow an investigation of whether the translations that I have developed are good ones or whether there are more natural ways to represent similar structures. While the particular algorithms that I have developed might not emerge with learning, the general techniques could be expected to manifest themselves. It is possible that some of these structures are not required to build real world models, but this seems unlikely based on the evidence of KL-ONE and some initial investigations with the bucket brigade. These structures are for computations that are useful in many domains and could be expected to play a role in most sophisticated models that are as powerful as KL-ONE. Since they are useful in KL-ONE, this suggests that they might be useful in other real world models.

A start has already been made in this direction. Goldberg [Goldberg, 83] and Holland [Holland, 85] have shown that the bucket brigade is capable of building up default hierarchies, using tags, using negative conditions as triggers, and limited sequencing (chaining). In addition, I would look for synchronization, more sophisticated uses of tags, more extensive sequencing, and in the context of knowledge representation, the formation of roles. Roles are more complex than "properties" for two reasons. First, they are two place relations rather than one place predicates, and second, relations between roles (DIFF links) are well defined. Of the other structures, it is possible that some are so central to every representation system that they should be "bootstrapped" into a learning system. That is, they should be provided from the beginning as a "macro" package and not required to be learned from the beginning every time.

References

Booker, Laiton (1982) "Intelligent Behavior as an Adaptation to the Task Environment", Ph. D. Dissertation (Computer and Communication Sciences) The University of Michigan, Ann Arbor, Michigan.

Brachman, Ronald J. (1978) "A Structural Paradigm for Representing Knowledge," Technical Report No. 3605, Bolt Beranek and Newman Inc., Cambridge, Ma.

Brachman, Ronald J. and Schmolze, James G. (1985), "An Overview of the KL-ONE Knowledge Representation System," Vol. 9, No. 2.

Fahlman, Scott E. (1979), NETL: A System for Representing and Using Real-World Knowledge, The MIT Press, Cambridge, Ma.

Forrest, Stephanie (1985), "A Study of Parallelism in The Classifier System and Its Application to Classification in KL-ONE Semantic Networks", Ph. D. Dissertation (Computer And Communication Sciences) The University of Michigan, Ann Arbor, Mi.

Goldberg, David (1983), Ph. D. Dissertation, The University of Michigan, Ann Arbor, Mi.

Holland, John H. (1975) Adaptation in Natural and Artificial Systems, The University of Michigan Press, Ann Arbor, Mi.

Holland, John H. (1980), "Adaptive Algorithms for Discovering and Using General Patterns in Growing Knowledge Bases", International Journal of Policy Analysis and Information Systems, Vol.4 No. 3.

Holland, John H. (1985), Personal Communication.

Lipkis, Thomas (1981), "A KL-ONE Classifier", Consul Note #5, USC/Information Sciences Institute, Marina del Rey, Ca.

Schmolze, James G. and Brachman, Ronald J. (1982) (editors) "Proceedings of the 1981 KL-ONE Workshop," Technical Report No. 4842, Bolt Beranek and Newman Inc., Cambridge, Ma.

Schmolze, James G. and Israel, David (1983), "KL-ONE: Semantics and Classification," in Sidner, C., et al., (editors) Technical Report No. 5421, Bolt Beranek and Newman Inc., Cambridge, Ma., pp. 27–39.

The Bucket Brigade is not Genetic

T. H. WESTERDALE

Abstract -- Unlike genetic reward schemes, bucket brigade schemes are subgoal reward schemes. Genetic schemes operating in parallel are here compared with a sequentially operating bucket brigade scheme. Sequential genetic schemes and parallel bucket brigade schemes are also examined in order to highlight the non-genetic nature of the bucket brigade.

I.   INTRODUCTION

The Bucket Brigade can be viewed as a class of apportionment of credit schemes for production systems. There is an essentially different class of schemes which we call genetic. Bucket Brigade schemes are subgoal reward schemes. Genetic schemes are not.

For concreteness, let us suppose the environment of each production system is a finite automaton, whose outputs are non-negative real numbers called payoffs. (To simplify our discussion, we are excluding negative payoff, but most of our conclusions will hold for negative payoff as well.) Each production's left hand side is a subset of the environment state set and each production's right hand side is a member of the environment's input alphabet. Associated with each production is a positive real number called that production's *availability*.

Probabilistic sequential selection systems are systems in which the following four steps take place each time unit: (1) The state of the environment is examined and those productions whose left hand sides contain this state form the eligibility set. (2) A member of the eligibility set is selected, probabilistically, each production in the set being selected with probability proportional to its availability. (3) This production then *fires*, which means merely that its right hand side is input into the environment causing an environment state transition and an output of payoff. (4) A reward scheme (or apportionment of credit scheme) examines the payoff and on its basis adjusts the availabilities of the various productions. Thus the availabilities are real numbers which are being continually changed by the reward scheme. Probabilistic sequential selection systems differ from one another in their differing reward schemes.

We assume that for any ordered pair of environment states there is a sequence of productions which will take us

45

from the first state to the second.

The average payoff per unit time is a reasonable measure of how well the system is doing. If the availabilities are held fixed, the system-environment complex becomes a finite state Markov Chain, and the average payoff per unit time (at equilibrium) is formally defined in the obvious way. As the availabilities change, the average payoff per unit time changes. Thus the average payoff per unit time can be thought of as a function of the availabilities. The object of the reward scheme is to change the availabilities so as to increase the average payoff per unit time.

The systems above have been simplified so as to more easily illustrate the points we wish to make. In any useful system the environment would output other symbols in addition to payoff, symbols which we could call ordinary output symbols. The left hand sides of the productions would then be sets of ordinary output symbols. A useful system would also contain some working memory (a "blackboard" or "message list") which could be examined and altered by the productions. In the above systems the working memory is regarded as part of the environment and instead of sets of output symbols we have sets of (Moore type) automaton states which produce those symbols. For illustrative purposes we have simplified the system by removing various parts and leaving only those parts on which the reward scheme operates.

In our systems the set of productions is fixed. We want to study the reward scheme, and allowing generation of new productions from old ones (E.g. [4]) will merely distract us.

II. GENETIC SYSTEMS WITH COMPLETE RECOMBINATION

At any given time, the production system can be thought of as a population of productions, the availability of a production giving the number of copies of that production in the population, or some fixed multiple of the number of copies. Thus the process of probabilistic selection of the production to fire can be thought of as randomly drawing productions from the population, until one is drawn that is in the eligibility set. In some systems the population is held explicitly and the availabilities are implicit whereas in others the availabilities are held explicitly and the population is implicit.

If the system is to be viewed as a population of productions, then of course after each production is successfully selected from the population it is tested on the environment with the environment in the state in which the previously selected production left it.

It is easier to analyse systems in which the result of a test of a production is independent of which productions were tested previously. Such systems are usually unrealistic, but if the system is viewed as a population of production strings, rather than of individual productions, then it is often realistic to view the test of a string as being independent of which strings were tested previously. Let us look at a population system of this kind. The system will consist of a population of production strings. The population will change over time. Time is viewed as divided into large units called generations. During a generation, every string in the population is tested against the environment and, as a result of the tests, the reward scheme determines the composition of the population in the next generation. A system of this kind we call a *string population system*. Let's examine such a system and give its reward scheme in detail. We shall call the system, *System A*. System A is a genetic system with complete recombination.

Begin with a set of productions, each with an availability. Let n and N be large integers with n much larger than N. The set of availabilities define a population of length n strings of productions (possibly with repeats) as follows. Let v be the sum of all the availabilities. For any length n string, the number of copies of that string in the population is $n^{-1}v^{1-n}$ times the product of the availabilities of its constituent productions. In each generation, the number of progeny of each string is given by testing the string and summing the payoff obtained during the test. To test a string one selects the first production in the string that is in the eligibility set, fires it, then moves on down the string until one finds the next production in the string that is now in the eligibility set, fires it etc. etc. until N productions have fired. We will not worry here about the few cases where one gets to the end of the string before N productions have fired.

We are assuming that during a generation, every string is tested against the environment. We are also assuming that there is an "initial state" of the environment and that when each string is tested the test always begins with the environment in the initial state, so that the results of a string test are independent of which strings were previously tested.

The formation of progeny is followed by complete recombination. In other words, each production's availability is incremented by the number of times that production occurs in the new progeny, and the next generation's population is formed from the availabilities just as the previous generation's population was. (In effect, the strings are broken into individual productions and these productions then re-combine at random to form a new population of length-n strings.)

We could have demanded that each string test begin with the environment in the state in which the last string left it, but if N and n are large then this demand will make hardly any difference to the test results. This is because the environment "forgets" what state it started in during a long test. For example, suppose there is one production whose left hand side is the set of all environment states and whose right hand side is a symbol which resets the environment to one particular state. Let's call this production the resetting production. Then during any string test, once the resetting production is encountered, the payoff for the rest of the test and the successive availability sets are independent of the state the environment was in when the test started. Thus each string has a value independent of which strings were tested previously, except for a usually small amount of payoff at the start of the test before the first occurrence of the resetting production. One can generalize these comments usefully to the case where there is no resetting production [6], but we will not do so formally here. The important thing to note is that except for a usually small initial segment, the sequence of successive eligibility sets would be independent of which strings were tested previously (provided n and N are large enough). Thus we do not lose anything important if we assume that each test begins with the environment in some initial state. So we can think of the tests in a generation as taking place sequentially or in parallel, it makes no difference.

Let the *value* of a string be the sum of the payoffs when the string is tested with the environment begun in the initial state. If there are x copies of a string ρ in the population, and if the value of ρ is y, then the number of progeny of ρ will be xy. If r is a production which occurs z times in the string ρ then zxy will be the contribution of the progeny of ρ to the increase in the availability of r. This is obvious, and we have only re-stated matters in this way to make it clear that we need not insist that x, y, and the activations are integers. The formalism makes perfect sense provided they are non negative real numbers. If the value of ρ is 0.038 then every copy of ρ will have 0.038 progeny. (But remember, we insist that activations, and hence x, are actually positive.)

Note that the behavior of System A can be thought of as a sequence of availability tuples. In any given generation, the population composition is given by the availabilities. Just as in the probabilistic sequential selection systems, the availabilities determine the average payoff per unit time (averaged over the tests of all the strings in the generation).

System A is deterministic. Given a tuple of availabilities it is completely determined what the next tuple of availabilities (in the next generation) will be. We will call two string population systems equivalent if they produce the same change in the availabilities, that is if given any tuple of availabilities, the next tuple of availabilities will be the same whichever system we are examining.

Actually we need a weaker notion of equivalence. We will also call two systems equivalent in several other circumstances. We will describe these circumstances informally, but will not give here a rigorous definition of equivalence.

Let the set of all possible tuples of availabilities be regarded as a subset of Euclidean space in the usual way. To each point in the subset corresponds an average payoff per unit time. System A defines for each point in the subset a vector giving the change in availabilities which its scheme would produce. Two systems are equivalent if at every point the change vector is the same for the two systems and the average payoff is also the same. We also call two systems equivalent if there is a positive scalar k such that at each point (1) the average payoff for the second system is k times that of the first, and (2) the change vector of the two systems aims in the same direction. So a system which was like System A but whose reward scheme always gave just half as many progeny would be equivalent to System A. If we define *normalizing* a vector as dividing it by the sum of its components, then condition (2) becomes "the normalized change vector of the two systems is the same."

For completeness I must mention a complication which will not be important in our discussion. We need to loosen condition (2) by normalizing the points in the space themselves. Normalizing a point in the space projects it onto the normalized hyperplane. (Its components can then be thought of as probabilities and it is of course these probabilities that we are really interested in.) If we take a change vector at a point, and think of the change vector as an arrow with its tail at that point, then we can normalize the point where its tail is and also normalize the point where its head is. The arrow between the two normalized points is a projection of the change vector onto the normalized hyperplane. We want condition (2) to say "the projected change vector of the two systems aims in the same direction", or "the normalized projected change vector of the two systems is the same". Sorry about this complication. It does make sense, but the details will not be important in our discussion.

Of course many schemes are probabilistic. Consider a system (System B) just like System A except that in each generation instead of its reward scheme giving progeny to all strings in the population, the reward scheme randomly selects just one string and gives only that string progeny (the same number of progeny System A would give it). Now the change in availabilities is probabilistic. At each point there are many possible change vectors, depending on which string is selected. When a system produces many possible change vectors at a point, we simply average them, weighting each possible change vector with the probability that it would represent the change. It is the average change vector that we then use in deciding system equivalence (or rather, the normalized projected average change vector). We call a scheme noisier the more the possible change vectors at a point differ from each other. So System B is equivalent to System A, though System B is much noisier.

Fisher's fundamental theorem of natural selection [1] [2] applies to Systems A and B so we know that for these systems the expectation of the change in the average payoff per unit time is non-negative. We call a system with this property *safe*. A safe system, then, is one in which at every point, the average change vector aims in a direction of non-decreasing average payoff. Clearly then, a system that is equivalent to a safe system is also safe.

Consider a system like System A except that the initial state (the state in which all string tests begin) is different from the initial state in System A. Technically, this new system would not be equivalent to System A, but if n and N are large enough it is nearly equivalent. In deciding system equivalence we will assume n and N are large enough. More precisely, we note that as n and N increase, a system's normalized projected average change vectors gradually change. At any point, the normalized projected average change vector approaches a limit vector as n and N approach infinity. It is this limit vector that we use as our normalized projected average change vector in deciding system equivalence. Thus the change in initial state produces a new system that is equivalent to System A. In fact, a system like A or B which begins each string test with the environment in the state the last string test left it is a system equivalent to A and B.

In all the systems discussed in this paper, a tuple of availabilities defines an average payoff per unit time, and the reward scheme defines, for each such tuple, an average change vector. This is true also in the probabilistic sequential selection systems. Thus we can compare any two of our systems and ask whether they are equivalent.

We ask if there is a reward scheme for a probabilistic sequential selection system that makes the system equivalent to Systems A and B. The natural candidate is System C defined by the following reward scheme: Reward every N productions which fire by incrementing the availabilities of these N productions by the sum of the payoffs over these N firings. But System C is not equivalent to System A. In the System A string tests, productions are skipped when they are not in the eligibility set. System A rewards these (increments their availabilities) whereas System C does not. To make C equivalent to A we must do something about rewarding the productions that are not in the eligibility set.

Equivalently we can instead penalize the various productions that are in the eligibility set. (See [5] for the formal details of the argument in the remainder of this section, including the effect of increasing string length.) The idea is that whenever production r is rewarded (has its availability incremented), the eligibility set R at the time r fired is penalized as follows. Let $\bar{S}$ be the sum of all availabilities and $\bar{R}$ the sum of the availabilities of productions in R. The *absolute probability* of r is the availability of r divided by $\bar{S}$. The *problbility of r relative to R* is the availability of r divided by $\bar{R}$. If the reward is x, the availability of r is first increased by x. Then the availabilities of all members of R are adjusted to bring $\bar{R}$ back down to what it was before the reward. The adjustment is done proportionally: i.e. the adjustments do not change the probabilities, relative to R, of the members of R. We call these adjustments penalties since they penalize a production for being eligible.

Let System C' be System C with this penalty scheme added. Then System C' is equivalent to Systems A and B.

In fact we can easily make this penalty scheme more sensible if we reward every time unit. The payoff in a time unit becomes the reward of the last N productions that fired (with corresponding penalties for the eligibility sets). This gives an equivalent, but more sensible scheme.

More sensibly, we can use an exponential weighting function, so that the reward of the production that fired z time units ago is c^z times the payoff. (c is a constant and $0 < c < 1$. Instead of assuming N large, we assume c is very close to 1.) This gives an equivalent scheme which is easy to implement because one only needs to keep for each production a count

$$\sum_{z=1}^{\infty} \chi(z) \cdot c^z$$

where $\chi(z)$ is 1 if the production fired z time units ago and 0 otherwise. A second count, called the production's eligibility count is also kept. This has the formula

$$\sum_{z=1}^{\infty} \frac{f(z)}{F(z)} \cdot c^z$$

where F(z) is the sum of the availabilities of the

51

productions that were in the eligibility set z time units ago and $f(z)$ is the availability of the given production if it was in the availability set z time units ago, and 0 otherwise. The count and the eligibility count are particularly easy to update. Each time unit, all productions are rewarded by the product of the current payoff times the difference between the count and the eligibility count. Call the probabilistic sequential selection system using this scheme System D.

System D is equivalent to System A. (Provided, as we said, we let n and N approach infinity and c approach 1.) Thus System D is safe (assymptotically safe as $c \rightarrow 1$) since System A is. Unfortunately a system using a bucket brigade scheme will not in general be safe, and it will not be equivalent to System A.

Since System D is equivalent to the genetic Systems A and B, we can call D also a genetic system. (Fisher's theorem says that a genetic system must be safe.) We can call the reward scheme of System D a genetic scheme for a probabilistic sequential selection system.

III. THE BUCKET BRIGADE

Genetic schemes like the scheme of System D form one class of reward schemes for probabilistic sequential selection systems. Another class is the class of bucket brigade schemes.

We shall examine the following bucket brigade scheme. Let C and K be constants, $0 < K \leqslant 1$, $0 < C \leqslant 1$. For each production the system holds two quantities, the availability and the *cash balance*. Productions are chosen from the eligibility set probabilistically on the basis of the availabilities. Each time unit, the production that fires pays proportion C of its cash balance to the production that fired in the previous time unit. The production that fired then has the current payoff added to its cash balance, and then its availability is increased by K times its cash balance. The members of the eligibility set are then penalized as in System C'. I know that Holland [4] employs a bidding system with the bucket brigade, but that system is much more difficult to analyze, so I shall use the probabilistic sequential selection system described above.

Let System E be a probabilistic sequential selection system using the above bucket brigade scheme. System E looks rather like System D. The most fundamental difference however is the following. In the bucket brigade, a production r is rewarded if it is followed by a production which is usually successful. In the genetic schemes of the last section, r is rewarded only if it is followed by a production which is successful this very time.

For this reason, a bucket brigade like the one in System E is not safe. Suppose the environment is such that the productions must come in triples, each triple being either hbc, dbe, or dfg. Suppose payoff is 10 for e, and zero for the other productions. Suppose hbc and dbe are equiprobable. The bucket brigade will pass b a cash payment of 5 on average, whereas f will get 0, so dfg will become less probable vis a vis dbe. Nevertheless, h gets passed 5 on average, whereas d gets passed less than 5, since it is sometimes followed by f, and g is broke.

The bucket brigade gives reward for achieving a subgoal. In the production system context, the subgoal is to put the environment in a state which will make eligible some production t. t says (via its left hand side) under what conditions it thinks it can convert the situation into one which will yield payoff. The subgoal is to provide those conditions. If r achieves that subgoal then (provided t is the production actually selected) r is rewarded. The amount of reward quite properly depends on how good t is, on how useful achieving that subgoal has been found in the past. This is the essence of a subgoal reward scheme.

Now of course it may happen that although t can usually achieve ultimate payoff, the particular state in which r happens to place the environment is one which (though included on the left hand side of t) t actually can never convert into ultimate payoff. That is, there is something slightly wrong with the left hand side of t. The subgoal (the left hand side of t) is badly formulated. In this case r will still be rewarded handsomely for achieving the subgoal since t usually does well. t of course will be mildly penalized for its indescriminateness since it will be in the eligibility set when it shouldn't be. Thus the effect of poor subgoal formulation can be to penalize the calling subroutine (t) for its poor formulation of the subgoal while rewarding the subroutine called (r) for the fact that it did what the calling routine asked. (I believe this is the correct analogy; the preceding production is the called subroutine and the following one is the calling subroutine.)

Now this is not what happens in the genetic schemes. In those schemes r would not get rewarded when followed by t. The called subroutine would only be rewarded for achieving a subgoal in the case where the ultimate result achieve payoff. Thus genetic schemes are not subgoal reward schemes.

At first sight it looks as if some reinterpretation of the system could reconcile the two approaches. Perhaps the bucket brigade will look genetic if an allele is something other than a single production. I can't prove that such a reconciling reinterpretation is impossible, but I'm rather convinced that the dichotomy between subgoal reward schemes and non subgoal reward schemes is too fundamental to permit of such a reinterpretation. What I shall do now is to try to highlight that dichotomy via one of the obvious ways of trying to make the bucket brigade look genetic. It will fail, and the way in which it fails will be illustrative.

Systems D and E are both probabilistic sequential selection systems. System D is equivalent to System A, a string population system. We now construct a string population system (System F) that is equivalent to System E. The ways in which systems F and A differ will be informative.

In System A, the composition of the population in a particular generation could be determined by examining a tuple of production availabilities. This will also be true in System F, though the way in which the availabilities determine the population composition will be different.

The trick in constructing System F is to find a way of explicitly stringing together those productions to which a bit of cash would be successively passed by the bucket brigade.

We form a population of length M strings of productions as follows by induction on M. If we have a population of length M strings, then the population of length M+1 strings is formed as follows. Take each length M string in turn. For each of these strings ρ, note the eligibility set R at the end of a test of it. (Each string test begins with the environment in the initial state.) For each r in R make 10000y copies of the string ρr , where y is the probability of r relative to R. Put these copies in the population of length M+1 strings.

We are interested in the case where M=N. These strings can be tested with no skipping.

In such a population of length N strings, any given production will occur many times, and may of course occur many times in the same string. For concreteness, let us call each production occurrence an *allele*, and let us say two alleles are of the same *type* if they are occurrences of the same production.

We are interested in two kinds of recombination. One is *ordinary complete recombination*, in which the strings are broken into individual alleles and these then recombine at random to form a new population of length N strings. The other is what we will call *crazy recombination*. In this the strings are broken into their individual alleles, but each

54

allele remembers the type of the allele that preceded it in its string. Then these alleles recombine at random, but an allele, in combining with others, must follow the same type of allele that it followed in its original string. The result of crazy recombination is a population of strings which represent the paths that cash may take in being passed by the bucket brigade.

System F works as follows. As described above, we use the availabilities to make a population of length N strings which can be tested with no skipping. The payoff for each of these strings is then determined, but it is not summed. Instead the system tests each string, noting for each time unit the eligibility set, the allele that fired, and the payoff received. The system then physically attaches the eligibility sets and payoffs to the alleles in the string, like clothes attached to a clothesline. To each allele in the string is attached the eligibility set from which that allele was selected and the payoff that arrived when that allele fired. Now we have a population of strings of alleles with each allele in each string having a number and a set attached to it. We now do crazy recombination, but the alleles carry their numbers and sets with them so that after recombination the strings still have numbers and sets hung along them. Now and only now are the payoffs for each string summed. Each production that occurs in the string is rewarded with this sum. For each allele in the string, the production of which that allele is an occurrence has the sum added to its availability; then the eligibility set attached to the allele is penalized in the usual way. The next generation population is formed from the new availabilities.

It is the crazy recombination that implements the bucket brigade notion that a production's reward depends on the production which follows and on how much reward that production achieved during some entirely different test. Though System F is equivalent to System E, it is a bit noisier because it is like a bucket brigade in which the cash is passed forward as well as backwards. Cash passed forwards, however, doesn't affect the biases and so doesn't affect equivalence in the sense in which we mean it.

System F doesn't look much like System A, but we can change System F slightly to improve matters. Carrying around the eligibility sets looks rather un-genetic. Instead of carrying around these sets and penalizing them we could carry around their complements and reward them. Equivalently (though with an increase in noise) we could carry around strings of productions selected from the complements and reward the productions in the strings much as we did in the original genetic scheme.

The idea here is to build a population of length n strings just as in System A. The system then marks on each string the first N alleles that will fire and attaches the appropriate payoffs to these alleles. Crazy recombination then proceeds as follows. (We can call this version *insane recombination*.) Each string is broken into segments, the breaks occurring at the N alleles. Each segment remembers the type of the allele that preceded it. The segments then recombine to form strings composed of N segments (the long tails which contain none of the N alleles are thrown away -- or, equivalently, they are attached to the ends of the new strings). In recombining, each segment must follow an allele of the same type as the one it followed before recombination. The new strings will now not all be of the same length. Each segment carries with it the payoff that was attached to its terminal production before the insane recombination. In each of the new strings the payoffs are summed and the sum gives the number of progeny of the new strings. Then the strings, including the new progeny are all broken apart into individual alleles and the total number of occurrences of a production (alleles) is its new availability. The population of the next generation is formed using the new availabilities. Thus this final breaking apart and formation of the next generation can be viewed as ordinary complete recombination.

So, beginning with the population thus formed, a generation consists of the following steps: (1) Mark on each string the N alleles which fire and attach to them the payoffs. (2) Do insane recombination, breaking at the N alleles, and having each segment carry its payoff with it. (3) Sum the payoffs on each of the new strings and produce the number of progeny given by the sum. (4) Do ordinary complete recombination.

Call the system using this scheme System G. System G is equivalent to System E.

The scheme of System G looks a bit like alternation of generations, but it has a dissatisfying artificiality. It is possible to remove the almost Lamarkian oddity of carrying payoffs attached to segments, but not in any particularly convincing fashion. Note that in step (3) we need not sum all the payoffs on a string. Equivalently we could just reward using one of those payoffs, or just a few. The payoffs we use need not be carried from before the insane recombination. They could be re-calculated afterwards. We need only take a re-combined string and run it, and then note one of the payoffs during the run. Actually it's not that trivial. The payoff we use must be one that arrives as the result of the firing of one of the N productions, not the other intervening productions.

Still, the scheme contains insane recombination and it is this that makes it inherently different from the genetic schemes.

56

IV.   FURTHER QUESTIONS

This raises several further questions which I find it difficult to answer.

(1)   The bucket brigade now begins to look rather silly. Is this just because of the bucket brigade version used here? Holland's bidding system bucket brigade is rather different. Perhaps it is the penalty scheme that is at fault. In genetic systems the penalty scheme preserves safety, but systems employing a bucket brigade aren't safe anyway. If the penalty scheme were removed we could go back to crazy recombination, but without the attached eligibility sets. This is less insane looking than insane recombination, but still not convincing. Nevertheless, we feel intuitively that subgoal reward is good. It certainly is a sensible way of combating scheme noise. So is it the genetic schemes that are the silly schemes? Or is there yet some way of viewing the bucket brigade so that it looks genetic?

(2)   If the genetic schemes and the bucket brigade are formally different, is there anyway a biological system analogue of the bucket brigade? One can imagine a crossing over rule that implements insane recombination. But there are two problems with this. One is that we have not merely insane recombination, but insane recombination alternating with ordinary recombination. Unfortunately the ordinary recombination is required in order to retain equivalence. It may be possible to remove the ordinary recombination and replace it with a phase which takes the strings and forgets which the special N productions are and then marks N new productions by determining which of the productions would be the first N to fire if that particular string were run. I don't think, though, that this change would retain equivalence.

(3)   In the more general context, do we see biological systems with subgoal reward? If so, then perhaps the bucket brigade has a sound biological basis, but it is merely that our usual population genetics formalism fails to capture that basis. For example, one might claim that a gene that does its own job (achieves a subgoal) more efficiently incurs lower cost, even if that job is useless for the current organism. The trouble is that if the job is converting metabolite A into B then the efficiency would probably give lower cost only if accumulation of B (or an equivalent) reduced gene expression. But then if B is not needed it will accumulate and reduce the cost of even an inefficient gene. Postulating leakage of B and other fiddles doesn't seem to help. The gene really only ends up rewarded when B is useful, and fiddling with cost only adjusts how much it is rewarded, but doesn't change the

basic fact that reward is much more when B is useful than when it isn't. So that doesn't seem to work.

A possibility not discussed in this paper is that in the bucket brigade the string of productions to which the cash is passed is the analog of a metabolic pathway in which each metabolite inhibits the expression of the gene responsible for the reaction that produces that metabolite [3]. Then each gene raises the expression rate (passes cash to) the gene preceding it in the pathway. In this view, the version of the bucket brigade described in this paper is wrong. In the correct version the cash is passed by a bucket brigade, the availabilities are adjusted by a genetic scheme that pays no attention to cash balances (unless you believe in Lamark), and the probabilities of the various productions firing are proportional to the products of the corresponding cash balances and availabilities. The biological analogue of these probabilities is then the prevalence of the various enzyme molecules rather than the prevalence of the various alleles.

Of course one can look for ecological models which use subgoal reward. This leads us into the quagmire of altruism, where current formalisms seem to me unhelpful.

It may be that examining some parasitic systems or symbiotic systems might be helpful. Parasites and endosymbionts must regulate their reproduction rate so as not to destroy a host. In effect they are passing cash to the host. But this is a situation where admittedly group selection is operating. Is it possible to regard two genes on the same chromosome as symbionts, regulating their reproduction rate to help each other? Perhaps we should explicitly implement group selection in our production systems (I don't believe this would be difficult) and let productions with various cash-passing schemes compete under such a group selection scheme.

V. CONCLUSIONS

It looks as if genetic systems are inherently different from systems employing the bucket brigade. We usually view genetic systems as population systems operating in parallel, whereas the bucket brigade operates essentially sequentially. This is a superficial difference. The essential difference appears to be that bucket brigade schemes are subgoal reward schemes, whereas genetic schemes are not.

REFERENCES

[1] J. F. Crow and M. Kimura, *An Introduction to Population Genetics Theory*. New York: Harper and Row, 1970.

[2] R. A. Fisher, *The Genetical Theory of Natural Selection*. New York: Dover, 1958.

[3] P. W. Hochachka and G. N. Somero, *Biochemical Adaptation* Princeton: Princeton University Press, 1984.

[4] J. H. Holland, "Escaping brittleness: the possibilities of general purpose learning algorithms applied to parallel rule based systems," to appear in *Machine Learning II* (R. S. Michalski, J. G. Carbonell, and T. M. Mitchell, eds.). Palo Alto: Tioga, 1984.

[5] T. H. Westerdale, "A Reward Scheme for Production Systems with Overlapping Conflict Sets," to appear in *IEEE Trans. Syst., Man, Cybern.*

[6] T. H. Westerdale, "An Automaton Decomposition for Learning System Environments," submitted.

GENETIC PLANS AND THE PROBABILISTIC LEARNING SYSTEM: SYNTHESIS AND RESULTS

Larry Rendell

Department of Computer Science,
University of Illinois at Urbana-Champaign,
1304 West Springfield Avenue, Urbana, Illinois 61801

ABSTRACT

This paper describes new conceptual and experimental results using the probabilistic learning system *PLS2*. PLS2 is designed for any task in which overall performance can be measured, and in which choice of task objects or operators influences performance. The system can manage incremental learning and noisy domains.

PLS2 learns in two ways. Its lower "perceptual" layer clusters data into economical cells or *regions* in augmented feature space. The upper "genetic" level of PLS2 selects successful regions (compressed *genes*) from multiple, parallel cases. Intermediate between performance data and task control structures, regions promote efficient and effective learning.

Novel aspects of PLS2 include compressed genotypes, credit localization and "population performance". Incipient principles of efficiency and effectiveness are suggested. Analysis of the system is confirmed by experiments demonstrating stability, efficiency, and effectiveness.

1. INTRODUCTION

The author's *probabilistic learning system PLS* is capable of efficient and effective generalization learning in many domains [Re 83a, Re 83d, Re 85a]. Unlike other systems [La 83, Mit 83, Mic 83a], PLS can manage noise, and learn incrementally. While it can be used for "single concept" learning, like the systems described in [Di 82], PLS has been developed and tested in the difficult domain of heuristic search, which requires not only noise management and incremental learning, but also removal of bias acquired during task performance [Re 83a]. The system can discover optimal evaluation functions (see Fig. 2). PLS has introduced some novel approaches, such as new kinds of clustering.[1]

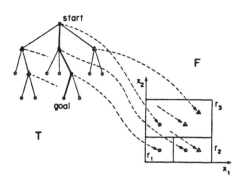

Figure 2. One use of PLS. In heuristic search, an object is a state, and its utility might be the probability of contributing to success (appearing on a solution path). E.g., for r_3, this probability is 1/3. Here the pair (r_3, p_3) is one of three *regions* which may be used to create a heuristic evaluation function. Region characteristics are determined by clustering.

---

1. See [Re 83a] for details and [Re 85a, Re 85b] for discussion of PLS's "conceptual clustering" [Mic 83b] which began in [Re 76, Re 77]. PLS "utility" of domain objects provides "category cohesiveness" [Me 85]. [Re 85c] introduces "higher dimensional" clustering which permits creation of structure. Appendix A summarizes some of these terms, which will be expanded in later sections of this paper.

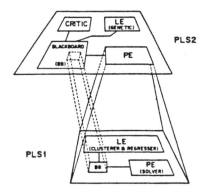

Figure 1. Layered learning system PLS2. The *perceptual* learning system PLS1 serves as the performance element (PE) of the *genetic* system PLS2. The PE of PLS1 is some task. PLS2 activates PLS1 with different knowledge structures ("cumulative region sets") which PLS2 continually improves. The basis for improvement is competition and credit localization.

Another successful approach to adaptation is *genetic algorithms* (GA's). Aside from their ability to discover global optima, GA's have several other important characteristics, including stability, efficiency, flexibility, and extensibility [Ho 75, Ho 81]. While the full behavior of genetic algorithms is not yet known in detail, certain characteristics have been established, and this approach compares very favorably with other methods of optimization [Be 80, Br 81, De 80]. Because of their performance and potential, GA's have been applied to various AI learning tasks [Re 83c, Sm 80, Sm 83].

In [Re 83c] a combination of the above two approaches was described: the doubly layered learning system *PLS2* (see Fig. 1).[2] PLS1, the lower level of PLS2, could be considered "perceptual"; it compresses goal-oriented information (task "utility") into a generalized, economical, and useful form ("regions" — see Figs. 2, 4). The upper layer is genetic, a competition of parallel knowledge structures. In [Re 83c], each of these components was argued to improve efficacy and efficiency.[3]

This paper extends and substantiates these claims, conceptually and empirically. The next section gives an example of a genetic algorithm which is oriented toward the current context. Section 3 describes the knowledge structure (regions) from two points of view: PLS1 and PLS2. Section 4 examines the synthesis of these two systems and considers some reasons for their efficiency. Sections 5 and 6 present and analyze the experimental results, which show the system to be stable, accurate, and efficient. The paper closes with a brief summary and a glossary of terms used in machine learning and genetic systems.

---

2. For the reader unfamiliar with learning system and other terminology, Appendix B provides brief explanations.

3. PLS2 is applicable to any domain for which features and "usefulness" or *utility* of objects can be defined [Re 83d]. An object can represent a physical entity or an operator over the set of entities. Domains can be simple (e.g. "single concept" learning), or complex (e.g. expert systems). State-space problems and games have been tested in [Re 83a, Re 83d]. The PLS approach is uniform and can be deterministic or probabilistic. The only real difficulty with a new domain is in constructing features which bear a smooth relationship to the utility (the system can evaluate and screen features presented to it).

2. GENETIC SYSTEMS: AN EXAMPLE

This section describes a simple GA, to introduce terminology and concepts, and to provide a basis for comparison with the more complex PLS2. The reader already familiar with GA's may wish to omit all but the last part of this section.

2.1. Optimization

Many problems can be regarded as function optimization. In an AI application, this may mean discovery of a good control structure for executing some task. The function to be optimized is then some measure of task success which we may call the *performance* μ. In the terminology of optimization, μ is the *objective function*. In the context of genetic systems, μ is the *fitness, payoff,* or *merit*.[4]

The merit μ depends on some control structure, the simplest example of which is a vector of *weights* $\mathbf{b} = (b_1, b_2, ..., b_n)$. Frequently the analytic form of $\mu(\mathbf{b})$ is not known, so exact methods cannot be used to optimize it (this is the case with most AI problems). But what often is available (at some cost) is the value of μ for a given control structure. In our example, let us suppose that μ can be obtained for any desired value of $\mathbf{b}$, by testing system performance. If μ is a well behaved, smooth function of $\mathbf{b}$, and if there is just one peak in the μ surface, then this *local optimum* is also a *global optimum*, which can be efficiently discovered using hill climbing techniques. However, the behavior of μ is often unknown, and μ may have numerous optima; in these cases a genetic adaptive algorithm is appropriate.

2.2. Genetic Algorithms

In a GA, a structure of interest, such as a weight vector $\mathbf{b}$, is called a *phenotype*. Fig. 3 shows a simple example with just two weights, b_1 and b_2. The phenotype is normally coded as a string of digits (usually bits) called the *genotype* B. A single digit is a *gene*; gene values are *alleles*. The position of a gene within the genotype is given by an index called the *locus*. Depending on the resolution desired, we might choose a greater or lesser number of sequential genes to code each b_i. If we consider 5 bits to be

---

4. μ might also be called the "utility", but we reserve this term for another kind of quality measure used by PLS1.

sufficient, the length of the genotype B will be L = 5n bits (see Fig. 3).

Instead of searching weight space directly for an optimal vector **b**, a GA searches gene space, which has dimensionality L (gene space is Hamming space if alleles are binary). A GA conducts this search in parallel, using a set of *individual* genotypes called a *population* or *gene pool*. By comparing the relative merits μ of individuals in a population, and by mating only the better individuals, a GA performs an informed search of gene space. This search is conducted iteratively, over repeated *generations*. In each new generation, there are three basic operations performed: (1) selection of parents, (2) generation of offspring, and (3) replacement of individuals. (1) and (2) have been given more attention. *Parent selection* is usually stochastic, weighted in favor of individuals having higher μ values. *Offspring generation* relies on *genetic operators* which modify parent genotypes. Two natural examples are mutation (which alters one allele), and crossover (which slices two genotypes at a common locus and exchanges segments — see Fig. 3).

POPULATION

Genotype B	Phenotype b	Merit μ
0001111110	(3, -2)	2.1
.		0.4
		0.8
0011011011	(6, -5)	1.7
.		0.7
		0.9
0010011100	(4, -4)	1.4

OFFSPRING

Parents B	Children B	Children b
0010 111110	0001011100	(2, -4)
crossover		
0001 011100	0010111110	(5, -2)

Figure 3. Simple genetic system. The upper part of this diagram shows a small population of just seven individuals. Here the set of characteristics (the *phenotype*) is a simple two element vector b. This is coded by the *genotype* B. Each individual is associated with its measured *merit* μ. On the basis of their μ values, pairs of individuals are stochastically chosen as parents for genetic recombination. Their genotypes are modified by crossover to produce two new offspring.

Because the more successful parents are selected for mating, and because limited opera-

tions are performed on them to produce offspring, the effect is a combination of knowledge retention and controlled search. Holland proved that, using binary alleles, the crossover operator, and parent selection proportional to μ, a GA is K^3 times more efficient than exhaustive search of gene space, where K is the population size [Ho 75, Ho 81]. Several empirical studies have verified the computational efficiency of GA's compared with alterative procedures for global optimization, and have discovered interesting properties of GA's, such as effects of varying K. For example, populations smaller than 50 can cause problems [Br 81, De 80].

2.3. Application in Heuristic Search

One AI use is search for solutions to problems, or for wins in games [Ni 80].[5] Here we wish to learn an evaluation function H as a combination of variables $x_1, x_2, ..., x_n$ called *attributes* or *features* (features are often used to describe states in search). In the simplest case, H is expressed as the linear combination $b_1 x_1 + b_2 x_2 + ... + b_n x_n = \mathbf{b.x}$, where the b_i are weights to be learned. We want to optimize the weight vector **b** according to some measure of the *performance* μ when H is used to control search.

A rational way to define μ (which we shall use throughout this paper) is related to the average number D of states or nodes developed in solving a set of problems. Suppose D is observed for a population of K heuristic functions H_i defined by weight vectors b_i. Since the performance improves with lower values of D, a good definition of the merit of H_i (i.e. of b_i) is the relative performance measure $\mu_i = \bar{D} / D_i$, where $\bar{D}$ is the average over the population, i.e. $\bar{D} = \sum D_j / K$. This expression of merit could be used to assess genotypes B representing weight vectors b_i, as depicted in Fig. 3.

Instead of this simple genetic approach, however, PLS2 employs unusual genotypes and operators, some of which relate to PLS1. In the remaining sections of this paper, we shall examine the advantages of the GA resulting from the combination of PLS1 with PLS2.

---

5. Notice that search takes place both at the level of the task domain (for good problem solutions), and at the level of the learning element (for a good control structure H).

3. PLS INFORMATION STRUCTURING: DUAL VIEWPOINT

The connection between PLS1 perceptual learning and PLS2 genetic adaptation is subtle and indirect. Basically PLS1 deals with *objects* **x** (which can be just about anything), and their relationships to task performance. Let us call the usefulness of an object **x** in some task domain its *utility* $u(\mathbf{x})$.

Since the number of objects is typically immense, even vast observation is incomplete, and generalization is required for prediction of u, given a previously unencountered **x**. A significant step in generalization is usually the expression of **x** as a vector of high-level, abstract features $x_1, x_2, ..., x_n$, so that **x** really represents not just one object, but rather a large number of similar objects (e.g. in a board game, **x** might be a vector of features such as piece advantage, center control, etc.). A further step in generalization is to *classify* or *categorize* **x**'s which are similar for current purposes.[6] Since the purpose is to succeed well in a task, PLS1 classifies **x**'s having similar utilities u.

Class formation can be accomplished in several ways, depending on the *model* assumed. If the task domain and features permit, objects having similar utilities may be *clustered* in feature space, as illustrated in Figs. 2 & 4, giving a "region set" R.[7] Another model is the linear combination $H = \mathbf{b}.\mathbf{f}$ of §2.

It is at this point that a GA like PLS2 can aid the learning process. Well performing **b**'s or R's may be selected according to their merit μ. Note that merit μ is an overall measure of the task performance, while utility u is a quality measure localized to individual objects.

The question now is what information structures to choose for representing knowledge about task utility. For many reasons, PLS incorporates the "region set" (Fig. 4), which represents domain knowledge by associating an object with its utility. We examine the region set from two points of view: as a PLS1 knowledge structure, and as a PLS2 genetic structure.

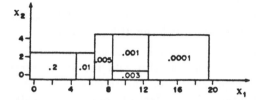

Figure 4. Dual interpretation of a region set R. A region set is a partition of feature space (here there are 6 regions). Points are clustered into regions according to their utility u in some task domain (e.g. u = probability of contributing to task success — see Fig. 2). Here the u values are shown inside the rectangles. A *region* R is the triple (r, u, e), where e is the error in u. The region set $\mathbf{R} = \{R\}$ serves both as the PLS1 knowledge structure and as the PLS2 genotype. In PLS1, R is a discrete (step) function expressing variation of utility u with features x_1. In PLS2, R is a compressed version of the detailed genotype illustrated in Fig. 5.

3.1. The Region as PLS1 Knowledge Structure

In a *feature space* representation, an object is a vector $\mathbf{x} = (x_1, x_2, ..., x_n)$.[8] In a problem or game, the basic object is the *state*, frequently expressed as a vector of features such as piece advantage, center control, mobility, etc.[9] Observations made during the course of even many problems or games normally cover just a fraction of feature space, and generalization is required for prediction.

In *generalization learning*, objects are abstracted to form *classes*, *categories*, or *concepts*. This may take the form of a partition of feature space, i.e. a set of mutually exhaustive local neighborhoods called *clusters* or *cells* [An 73, Di 82]. Since the goal of clustering in PLS is to aid task performance, the basis for generalization is some measure of the worth, quality, or

---

6. Here *to classify* means *to form* classes, categories, or concepts. This is difficult to automate.

7. PLS1 initiated what has become known as conceptual clustering — where not just feature values are considered, but also predetermined forms of classes (e.g. rectangles), and the whole data environment (e.g. utility). See [Re 76, Re 77, Re 83a, Re 85a, Re 85b], and also Appendix A.

8. Feature spaces are sometimes avoided because they cannot easily express structure. However, alternative representations, as normally used, are also deficient for realistic generalization learning. A new scheme mechanizes of a very difficult inductive problem: feature *formation* [Re 83d, Re 85c].

9. The object or event could just as well be an operator to be applied to a state, or a state-operator pair. See [Re 83d].

utility of a state or cell, relative to the task. One measure of utility is the probability of contributing to a solution or win. In Figs. 2, 4, probability classes are rectangular cells (for economy). The leftmost rectangle r has probability $u = 0.2.$[10] The rectangle r is a category generalizing the conditions under which the utility u applies.

In PLS, a rectangle is associated not just with its utility u, but also with the utility *error* e. This expression e of uncertainty in u allows quantification of the effect of noise and provides an informed and concise means for weighting various contributions to the value of u during learning. The triple $R = (r, u, e)$, called a *region*, is the main knowledge structure for PLS1. A set $R = \{R\}$ of regions defines a partition in *augmented* feature space.

R may be used directly as a (discrete) *evaluation* or *heuristic* function $H = u(r)$ to assess state $x \in r$ in search. For example, in Fig. 4, there are six regions, which differentiate states into six utility classes. Instead of forming a discrete heuristic, R may be used indirectly, as data for determining the *weight vector* b in a smooth evaluation function $H = b.x$ (employing curve fitting techniques). We shall return to these algorithmic aspects of PLS in §4.

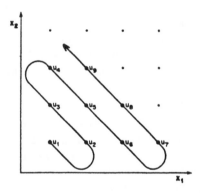

Figure 5. Definition of maximally detailed genotype U. If the number of points in feature space is finite and a value of the utility is associated with each point, complete information can be captured in a detailed genotype U of concatenated utilities $u_1 u_2 \ldots u_L$. Coordinates could be linearly ordered as shown here for the two dimensional case. U is an fully expanded genotype corresponding to the compressed version of Fig. 4.

---

10. This could be expressed in other ways. The production rule form is $r \to u$. Using logic, r is represented: $(0 \le x_1 \le 4) \cap (0 \le x_2 \le 2)$.

3.2. The Region Set as Compressed and Unrestricted PLS2 Genotype

Now let us examine these information structures from the genetic viewpoint. The weight vector b of evaluation function H could be considered a GA phenotype. What might the genotype be? One choice, a simple one, was described in §2 and illustrated in Fig. 3: here the genotype B is just a binary coding of b. A different possibility is one that captures exhaustive information about the relationship between utility u and feature vector x (see Fig. 5). In this case, the gene would be (x, u). If the number of genes is finite, they can be indexed and concatenated, to give a very detailed genotype U, which becomes a string of values $u_1 u_2 \ldots u_L$ coding the entire utility surface in augmented feature space.

This genotype U is unusual in some important ways. Let us compare it with the earlier example B of §2 (Fig. 3). B is simply a binary form of weight vector b. One obvious difference between B and U is that U is more verbose than B. This redundancy aspect will be considered shortly. The other important difference between B and U is that alleles within B may well interact (to express feature nonlinearity), but alleles u_i within U *cannot* interact (since the u_i express an absolute property of feature vector x, i.e. its utility for some task). As explained in the next section, this freedom from gene interdependence permits localization of credit.[11]

The detailed genotype U codes the utility surface, which may be very irregular at worst, or very smooth at best. This surface may be locally well behaved (it may vary slowly in some volumes of feature space). In cases of local regularity, portions of U are redundant. As shown in Fig. 5, PLS2 compresses the genotype U, into the region set R (examined in §3.1 from the PLS1 viewpoint). In PLS2, a single region $R = (r, u, e)$ is a *set* of genes, the whole having just one allele u (we disregard the genetic coding of e). Unlike standard genotypes, which have a stationary locus for each gene and a fixed number of genes, the region set has no explicit loci, but rather a

---

11. While one of the strengths of a GA is its ability to manage interaction of variables (by "co-adapting" alleles), PLS2 achieves efficient and concise knowledge representation and acquisition by flexible gene compression, and by certain other methods examined later in this paper.

variable number of elements (regions), each representing a variable number of genes. A region compresses gene sets having similar utility according to current knowledge.

4. KNOWLEDGE ACQUISITION: SYNERGIC LEARNING ALGORITHMS

In this section we examine how R is used to provide barely adequate information about the utility surface. This compact representation results in economy of both space and time, and in effective learning. Some reasons for this power are considered.

The ultimate purpose of PLS is to discover utility classes in the form of a region set R. This knowledge structure controls the primary task: for example, in heuristic search, $R = \{R\} = \{(r, u, e)\}$ defines a discrete evaluation function $H(r) = u$.

The ideal R would be perfectly *accurate* and maximally *compressed*. Accuracy of utility u determines the quality of task performance. Appropriate compression of R characterizes the task domain concisely but adequately (see Figs. 3, 4), saving storage and time, both during task performance and during learning.

These goals of accuracy and economy are approached by the doubly layered learning system PLS2 (Fig. 1). PLS1 and PLS2 combine to become effective rules for generalization (induction), specialization (differentiation), and reorganization. The two layers support each other in various ways: for example PLS2 stabilizes the perceptual system PLS1, and PLS1 maintains genotype diversity of the genetic system PLS2. In the following we consider details, first from the standpoint of PLS1, then from the perspective of PLS2.

4.1. PLS1 Revision and Differentiation

Even without a genetic component, PLS1 is a flexible learning system which can be employed in noisy domains requiring incremental learning. It can be used for simple concept learning like the systems in [Di 82], but most experiments have involved state space problem solving and game playing.[12] Here we examine PLS in the context of

these difficult tasks.

As described in §3.1, the main PLS1 knowledge structure is the region set $R = \{(r, u, e)\}$. Intermediate between basic data obtained during search, and a general heuristic used to control search, R defines a feature space augmented and partitioned by u and e. Because R is improved incrementally, it is called the *cumulative region set*. PLS1 repeatedly performs two basic operations on R. One operation is correction or *revision* (of utility u and error e), and the other is specialization, differentiation, or *refinement* (of feature space cells r). These operators are detailed in [Re 83a, Re 83d]; here we simply outline their effects and note their limitations.

Revision of u and e. For an established region $R = (r, u, e) \in R$, PLS1 is able to modify u and to decrease e by using new data. This is accomplished in a rough fashion, by comparing established values within all rectangles r with fresh values within the same r. It is difficult or impossible to learn the "true" values of u, since data are acquired during performance of hard tasks, and these data are biased in unknown ways because of nontrivial search.

Refinement of R. Alternately performing then learning, PLS1 acquires more and more detail about the nature of variation of utility u with features. This information accumulates in the region set $R = \{R\} = \{(r, u, e)\}$, where the primary effect of clustering u is increasing resolution of R. The number, sizes, and shapes of rectangles in R reflect current knowledge resolution. As this differentiation continues in successive iterations of PLS1, attention focuses on more useful parts of feature space, and heuristic power improves.

Unfortunately, so does the likelihood of error. Further, errors are difficult to quantify and hard to localize to individual regions.

In brief, while the incremental learning of PLS1 is powerful enough to learn locally optimal heuristics under certain conditions, and while PLS1 feedback is good enough to control and correct mild errors, the feedback can become unstable in unfavorable situations: instead of being corrected, errors can become more pronounced. Moreover, PLS1 is sensitive to parameter settings (see Appendix B). The system needs support.

---

12. These experiments have led to unique results such as discovery of locally optimal evaluation functions (see [Re 83a, Re 83d]).

4.2. PLS2 Genetic Operators

Qualities missing in PLS1 can be provided by PLS2. As §4.1 concluded, PLS1, with its single region set, cannot discover accurate values of utilities u. PLS2, however, maintains an entire population of region sets, which means that several regions in all cover any given feature space volume. The availability of comparable regions ultimately permits greater accuracy in u, and brings other benefits.

As §3.2 explained, a PLS2 genotype is the region set $R = \{R\}$, and each region $R = (r, u, e)$ $\in R$ is a compressed gene whose allele is the utility u. Details of an early version of PLS2 are given in [Re 83c]. Those algorithms have been improved; the time complexity of the operators in recent program implementations is linear with population size K. The following discussion outlines the overall effects and properties of the various genetic operators (compare to the more usual GA of §2).

K-sexual mating is the operator analogous to crossover. Consider a population $\{R\}$ of K different region sets R. Each set is composed of a number of regions R which together cover feature space. A new region set R' is formed by selecting individual regions (one at a time) from parents R, with probability proportional to merit μ (merit is the performance of R defined at the end of §2). Selection of regions from the whole population of region sets continues until the feature space cover is approximately the average cover of the parents. This creates the offspring region set R' which is generally not a partition.

Gene reorganization. For economy of storage and time, the offspring region set R' is repartitioned so that regions do not overlap in feature space.

Controlled mutation. Standard mutation operators alter an allele randomly. In contrast, the PLS2 operator analogous to mutation changes an allele according to evidence arising in the task domain. The controlled mutation operator for a region set $R = \{(r, u, e)\}$ is the utility revision operator of PLS1. As described in §4.1, PLS1 modifies the utility u for each feature space cell r.

Genotype expansion. This operator is also provided by PLS1. Recall the discussion of §3.2 about the economy resulting from compressing genes (utility-feature vectors) into a region

set R. The refinement operator was described in §4.1. This feature space refinement amounts to an expansion of the genotype R, and is carried out when data warrant the discrimination.

Both controlled mutation and genotype expansion promote genotype diversity. Thus PLS1 helps PLS2 to avoid premature convergence, a typical GA problem [Br 81, Ma 84].

4.3. Effectiveness and Efficiency

The power of PLS2 may be traced to certain aspects of the perceptual and genetic algorithms just outlined. Some existing and emerging principles of effective and efficient learning are briefly discussed below (see also [Re 85a, Re 85b, Re 85c]).

Credit localization. The selection of regions for K-sexual mating may use a single merit value μ for each region R within a given set R. However, the value of μ can just as well be localized to single regions within R, by comparing R with similar regions in other sets. Since regions estimate an absolute quantity (task-related utility) in their own volume of feature space, they are independent of each other. Thus credit and blame may be assigned to feature space cells (i.e. to gene sequences).

Assignment of credit to individual regions within a cumulative set R is straightforward, but it would be difficult to do directly in the final evaluation function H, since the components of H, while appropriate for performance, omit information relevant to learning (compare Figs. 2, 4).[13]

Knowledge mediation. Successful systems tend to employ information structures which *mediate* data objects and the ultimate knowledge form. These mediating structures include means to record growing assurance of tentative hypotheses.

When used in heuristic search, the PLS region set mediates large numbers of states and a

---

13. There are various possibilities for the evaluation function H, but all contain less useful information than their determinant, the region set R. The simplest heuristic used in [Re 83a, Re 83d] is $H = b.f$, where b is a vector of weights for the feature vector f. (This linear combination is used exclusively in experiments to be described in §5.) The value of b is found using regions as data in linear regression [Re 83a, Re 83b].

very concise evaluation function H. Retention and continual improvement of this mediating structure relieves the credit assignment problem. This view is unlike that of [Di 81, p. 14, Di 82]: learning systems often attempt to improve the control structure itself, whereas PLS acquires knowledge efficiently in an appropriate structure, and utilizes this knowledge by compressing it only temporarily for performance. In other words, PLS does not directly search rule space for a good H, but rather searches for good cumulative regions from which H is constructed.

Full but controlled use of every datum. Samuel's checker player permitted each state encountered to influence the heuristic H, and at the same time no one datum could overwhelm the system. The learning was stochastic: both conservative and economic. In this respect PLS2 is similar (although more automated).

Schemata in learning systems and genetic algorithms. A related efficiency in both Samuel's systems and PLS is like the schemata concept in a GA. In a GA, a single individual, coded as a genotype (a string of digits), supports not only itself, but also all its substrings. Similarly, a single state arising in heuristic search contains information about every feature used to describe it. Thus each state can be used to appraise and weight each feature. (The effect is more pronounced when a state is described in more elementary terms, and combinations of primitive descriptors are assessed — see [Re 85c]).

5. EXPERIMENT AND ANALYSIS

PLS2 is designed to work in a changing environment of increasingly difficult problems. This section describes experimental evidence of effective and efficient learning.

5.1. Experimental Conditions

Tame features. The features used for these experiments were the four of [Re 83a]. The relationship between utility and these features is fairly smooth, so the full capability of a GA is not tested, although the environment was dynamic.

Definition of merit μ. As §4 described, PLS2 choses regions from successful cumulative sets and recombines them into improved sets. For the experiments reported here, the selection criterion was the global merit μ, i.e. the performance of a whole region *set*, without localization of credit to individual regions. This measure μ was the average number of nodes developed D in a training sample of 8 fifteen puzzles, divided into the mean of all such averages in a population of K sets, i.e. $\mu = \bar{D}/D$, where $\bar{D}$ is the average over the population ($\bar{D} = \sum D_j / K$).

Changing environment. For these experiments, successive rounds of training were repeated in incremental learning over several iterations or *generations*. The environment was altered in successive generations; it was specified as problem *difficulty* or depth d (defined as the number of moves from the goal in sample problems). As a sequence of specifications of problem difficulty, this becomes a training *difficulty vector* $\mathbf{d} = (d_1, d_2, ..., d_n)$.

Here $\mathbf{d}$ was static, one known to be a good progression, based on previous experience with user training [Co 84].[14] In these experiments, $\mathbf{d}$ was always $(8, 14, 22, 50, \#, \#, ...)$. An integer means random production of training problems subject to this difficulty constraint, while "#" demands production of fully random training instances.

5.2. Discussion

Before we examine the experiments themselves let us consider potential differences between PLS1 and PLS2 in terms of their effectiveness and efficiency. We also need a criterion for assessing differences between the two systems.

Vulnerability of PLS1. With population size $K = 1$, PLS2 degenerates to the simpler PLS1. In this case, static training can result in utter failure, since the process is stochastic and various things can go wrong (see Appendix B). The worst is failure to solve any problems in some generation, and consequent absence of any new information. If the control structure H is this poor, it will not improve unless the fact is

---

14. PLS and similar systems for problems and games are sometimes neither fully supervised nor fully unsupervised. The original PLS1 was intermediate in this respect. Training problems were selected by a human, but from each training instance, a multitude of individual nodes for learning are generated by the system. Each node can be considered a separate example for concept learning [Re 83d]. [Co 84] describes experiments with an automated trainer.

detected and problem difficulty is reduced (i.e. dynamic training is needed).

Even without this catastrophe, PLS1 performs with varying degrees of success depending on the sophistication of its training and other factors (explained in Appendix B). With minimal human guidance, PLS1 always achieves a good evaluation function H, although not always an optimal one. With static training, PLS1 succeeds reasonably about half the time.

Stability of PLS2. In contrast, one would expect PLS2 to have a much better success rate. Since PLS1 is here being run in parallel (Fig. 1), and since PLS2 should reject hopeless cases (their μ's are small), a complete catastrophe (*all* H's failing) should occur with probability $p \leq q^K$, where q is the probability of PLS1 failure and K is population size. If q is even as large as one half, but K is 7 or more, the probability p of catastrophe is less than 0.01.

Cost versus benefit: a measure. Failure plays a part in costs, so PLS2 may have an advantage. The ultimate criterion for system quality is cost effectiveness: is PLS2 worth its extra complexity? Since the main cost is in task performance (here solving), the number of nodes developed D to attain some performance is a good measure of the expense.

If training results in catastrophic failure, however, all effort is wasted, so a better measure is the expected cost D/p, where p is the probability of success. For example, if D = 500 for viable control structures, but the probability of finding solutions is only ½, then the average cost of *useful* information is 500 / ½ = 1000.

To extend this argument, probability p depends on what is considered a success. Is success the discovery of a perfect evaluation function H, or is performance satisfactory if D departs from optimal by no more than 25%?

5.3. Results

Table I shows performances and costs with various values of K. Here p is estimated using roughly 36 trials of PLS1 in a PLS2 context (if K = 1, 36 distinct runs; if K = 2, 18 runs; etc.). Since variances in D are high, performance tests were made over a random sample of 50 puzzles. This typically gives 95% confidence of $D \pm 40$.

Accuracy of learning. Let us first compare results of PLS1 versus PLS2 for four different success criteria. We consider the learning to be successful if the resulting heuristic H approaches optimal quality within a given margin (of 100%, 50%, 25%, and 10%).

Columns two to five in the table (the second group) show the proportion of H's giving performance D within a specified percentage of the best known D (the best D is around 350 nodes for the four features used). For example, the last row of the table shows that, of the 36 individual control structures H tested in (two different) populations of size 19, all 36 were within 100% of optimal D (column two). This means that all developed no more than 700 nodes before a solution was found. Similarly, column five in the last row shows that 0.21 of the 36 H's, or 8 of them, were within 10%, i.e. required no more than 385 nodes developed.

Cost of accuracy. Columns ten and eleven (the two rightmost columns of the fourth group) show the estimated costs of achieving performance within 100% and within 10% of optimum, respectively. The values are based on the expected total number of nodes required (i.e. D/p), with adjustments in favor of PLS1 for extra PLS2 overhead. (The unit is one thousand nodes developed.) As K increases, the cost of a given accuracy first increases. Nevertheless, with just moderate K values, the genetic system becomes cheaper, particularly for an accuracy of 10%.

TABLE 1. COSTS and PERFORMANCES at GENERATION 5.

Pop. Size K	Proportion Satisfying Success Criterion (proximity to optimal D)				Mean Nodes Developed (Random Sample of 50)			Cost Per Individual H [10^3 nodes]	Expected Cost of One H Within		H Performance Rough Estimate (for cost=4×10^5)
	100%	50%	25%	10%	Avg $\bar{D}$	Best D_b	Pop. D_p		100%	10%	
1	.47	.28	.12	.04	>>654	375	–	12.5	26.6	312	31%
4	.69	.31	.19	.03	573	381	406	17.4	25.2	580	30%
7	.82	.43	.31	.08	531	377	395	18.1	22.7	226	20%
12	1.00	.53	.39	.11	507	384	397	18.7	18.7	170	12%
15	1.00	.69	.61	.14	450	370	390	19.2	19.2	137	9%
19	1.00	.71	.63	.21	453	348	367	19.7	19.7	94	7%

The expected cost benefit is not the only advantage of PLS2.

Average performance, best performance, and population performance. Consider now the third group of columns in Table I. The sixth column gives the *average* $\bar{D}$ for all H's in the sample (of 36). The seventh column gives D_b for the *best* H_b in the sample. These two measures, *average* and *best* performance, are often used in assessing genetic systems [Br 81]. The eighth column, however, is unusual; it indicates the *population performance* D_p resulting when all regions from every set in the population are used together in a regression to determine H_p. This is sensible because regions are independent and estimate the utility, an absolute quantity ([Re 83c], cf [Br 81, Ho 75]).

Several trends are apparent in Table I. First, whether the criterion is 100%, 50%, 25%, or 10% of optimum (columns 2-5), the proportion of good H's increases steadily as population size K rises. Similarly, average, best, and population performance measures $\bar{D}$, D_b and D_p (columns 6-8) also improve with K. Perhaps most important is that the population performance D_p is so reliably close to best, even with these low K values. This means that the whole population of regions can be used (for H_p) without independent verification of performance. In contrast, individual H's would require additional testing to discover the best (column 7), and the other alternative, any H, is likely not as good as H_p (columns 6 and 8). Furthermore, the entire population of regions can become an accurate source of massive data for determining an evaluation function capturing feature interaction [Re 83b].

This accuracy advantage of PLS2 is illustrated in the final column of the table, where, for a constant cost, rough estimates are given, of the expected error in population performance D_p relative to the optimal value.

It is interesting that such small populations improve performance markedly; usually population sizes are 50 or more.

6. EFFICIENCY AND CAPABILITY

Based on these empirical observations for PLS2, on other comparisons for PLS1, and on various conceptual differences, general properties of three competing methods can be compared: PLS1, PLS2, and traditional optimization. In [Re 81], PLS1 was found considerably more efficient than standard optimization, and the suggestion was made that PLS1 made better use of available information. By studying such behaviors and underlying reasons, we should eventually identify principles of efficient learning. Some aspects are considered below.

Traditional optimization versus PLS1. First, let us consider efficiency of search for an optimal weight vector **b** in the evaluation function $H = \mathbf{b.f}$. One good optimization method is *response surface fitting* (RSF). It can discover a local optimum in weight space by measuring and regressing the response (here number of nodes developed D) for various values of **b**. RSF utilizes just a single quantity (i.e. D) for every problem solved. This seems like a small amount of information to extract from an entire search, since a typical one may develop hundreds or thousands of nodes, each possibly containing relevant information. In contrast to this traditional statistical approach, PLS1, like [Sa 63, Sa 67], uncovers knowledge about *every* feature from *every* node (see §4.3). PLS1, then, might be expected to be more efficient than RSF. Experiments verify this [Re 81].

Traditional optimization versus PLS2. As shown in §5, PLS2 is more efficient still. We can compare it, too, with RSF. The accuracy of RSF is known to improve with $\sqrt{N}$, where N is the number of data (here the number of of problems solved). As a first approximation, a parallel method like PLS2 should also cause accuracy to increase with the square root of the number of data, although the data are now regions instead of D values. If roughly the same number of regions is present in each individual set R of a population of size K, accuracy must therefore improve as $\sqrt{K}$. Since each of these K structures requires N problems in training, the accuracy of PLS2 should increase as $\sqrt{N}$, like RSF.

Obviously, though, PLS2 involves much more than blind parallelism: a genetic algorithm extracts accurate knowledge and dismisses incorrect (unfit) information. While it is impossible for PLS1 alone, PLS2 can refine merit by localizing credit to individual regions [Re 83c]. Planned experiments with this should show further increases in efficiency since the additional cost is small. Another inexpensive improvement will attempt to reward good regions by decreasing their estimated errors. Even without these refinements, PLS2 retains meritorious regions (§4), and should exhibit accuracy improvement better than $\sqrt{N}$. Table I suggests this.

PLS2 versus PLS2. As discussed in §4.1 and Appendix B, PLS1 is limited, necessitating human tuning for optimum performance. In contrast, the second layer learning system PLS2 requires little human intervention. The main reason is that PLS2 stabilizes knowledge automatically, by comparing region sets and dismissing aberrant ones. Accurate cumulative sets have a longer lifetime.

This ability to discriminate merit and retain successful data will likely be accentuated with the localization of credit to individual regions (see §4.2). Another improvement is to alter dynamically the error of a region (estimated by PLS1) as a function of its merit (found by PLS2). This will have the effect of protecting a good region from imperfect PLS1 utility revision; once some parallel PLS1 has succeeded in discovering an accurate value, it will be more immune to damage. A *fit* region will have a very long lifespan.

Inherent differences in capability. RSF, PLS1, and PLS2 can be characterized differently. From the standpoint of time costs: given a challenging requirement such as the location of a local optimum within 10%, the ordering of these methods in terms of efficiency is RSF ≤ PLS1 ≤ PLS2. In terms of *capability*, the same relationship holds. RSF cannot handle feature interactions without a more complex model (which would increase its cost drastically). PLS1, on the other hand, can provide some performance improvement using piecewise linearity, with little additional cost [Re 83b]. PLS2 is more robust than PLS1. While the original system is somewhat sensitive to training and parameters, PLS2 provides stability using competition to overcome deficiencies, obviate tuning, and increase accuracy, all at once. PLS2 buffers inadequacies inherent in PLS1. Moreover, PLS2, being genetically based, may be able to handle highly interacting features, and discover *global* optima [Re 83c]. This is very costly with RSF and seems infeasible with PLS1 alone.

7. SUMMARY AND CONCLUSIONS

PLS2 is a general learning system [Re 83a, Re 83d]. Given a set of user-defined features and some measure of the *utility* (e.g. probability of success in task performance), PLS2 forms and refines an appropriate knowledge structure, the *cumulative region set* R, relating utility to feature values, and permitting noise management. This economical and flexible structure *mediates* data objects and abstract heuristic knowledge.

Since individual regions of the cumulative set R are independent of one another, both credit localization and feature interaction are possible simultaneously. Separating the task control structure H from the main store of knowledge R allows straightforward credit assignment to this *determinant* R of H, while H itself may incorporate feature nonlinearities without being responsible for them.

A concise and adequate embodiment of current heuristic knowledge, the cumulative region set R was originally used in the learning system PLS1 [Re 83a]. PLS1 is the only system shown to discover locally optimal evaluation functions in an AI context. Clearly superior to PLS1, its genetic successor PLS2 has been shown to be more stable, more accurate, more efficient, and more convenient. PLS2 employs an unusual genetic algorithm having the cumulative set R as a compressed genotype. PLS2 extends PLS1's limited operations of revision (controlled mutation) and differentiation (genotype expansion), to include generalization and other rules (K-sexual mating and genotype reorganization). Credit may be localized to individual gene sequences.

These improvements may be viewed as effecting greater efficiency or as allowing greater capability. Compared with a traditional method of optimization, PLS1 is more efficient [Re 85a], but PLS2 does even better. Given a required accuracy, PLS2 locates an optimum with lower expected cost. In terms of capability, PLS2 insulates the system from inherent inadequacies and sensitivities of PLS1. PLS2 is much more stable and can use the whole population of regions reliably to create a highly informed heuristic (this *population performance* is not meaningful in standard genetic systems). This availability of massive data has important implications for feature interaction [Re 83b].

Additional refinements of PLS2 may further increase efficiency and power. These include rewarding meritorious regions so they become immune to damage. Future experiments will investigate nonlinear capability, ability to discover global optima, and efficiency and effectiveness of localized credit assignment.

This paper has quantitatively affirmed some principles believed to improve efficiency and effectiveness of learning (e.g. credit localization). The paper has also considered some simple but little explored ideas for realizing these capabilities (e.g. full but controlled use of each datum).

REFERENCES

[An 73] Anderberg, M.R., *Cluster Analysis for Applications*, Academic Press, 1973.

[Be 80] Bethke, A.D., *Genetic algorithms as function optimizers*, Ph.D. Thesis, University of Michigan, 1980.

[Br 81] Brindle, A., Genetic algorithms for function optimization, C.S. Department Report TR81-2 (PhD Dissertation), University of Alberta, 1981.

[Bu 78] Buchanan, B.G., Johnson, C.R., Mitchell, T.M., and Smith, R.G., Models of learning systems, in Belzer, J. (Ed.), *Encyclopedia of Computer Science and Technology 11* (1978), 24-51.

[Co 84] Coles, D. and Rendell, L.A., Some issues in training learning systems and an autonomous design, *Proc. Fifth Biennial Conference of the Canadian Society for Computational Studies of Intelligence*, 1984.

[De 80] DeJong, K.A., Adaptive system design: A genetic approach, *IEEE Transactions on Systems, Man, and Cybernetics SMC-10*, (1980), 566-574.

[Di 81] Dietterich, T.G. and Buchanan, B.G., The role of the critic in learning systems, Stanford University Report STAN-CS-81-891, 1981.

[Di 82] Dietterich, T.G., London, B., Clarkson, K., and Dromey, G., Learning and inductive inference, STAN-CS-82-913, Stanford University, also Chapter XIV of *The Handbook of Artificial Intelligence*, Cohen, P.R., and Feigenbaum, E.A. (Ed.), Kaufmann, 1982.

[Ho 75] Holland, J.H., *Adaptation in Natural and Artificial Systems*, University of Michigan Press, 1975.

[Ho 80] Holland, J.H., Adaptive algorithms for discovering and using general patterns in growing knowledge bases, *Intl. Journal on Policy Analysis and Information Systems 4*, 2 (1980), 217-240.

[Ho 81] Holland, J.H., Genetic algorithms and adaptation, *Proc. NATO Adv. Res. Inst. Adaptive Control of Ill-defined Systems*, 1981.

[Ho 83] Holland, J.H., Escaping brittleness, *Proc. Second International Machine Learning Workshop*, 1983, 92-95.

[La 83] Langley, P., Bradshaw, G.L., and Simon, H.A., Rediscovering chemistry with the Bacon system, in Michalski, R.S., Carbonell, J.G., and Mitchell, T.M. (Ed.), *Machine Learning: An Artificial Intelligence Approach*, Tioga, 1983, 307-329.

[Ma 84] Mauldin, M.L., Maintaining diversity in genetic search, *Proc. Fourth National Conference on Artificial Intelligence*, 1984, 247-250.

[Me 85] Medin, D.L. and Wattenmaker, W.D., Category cohesiveness, theories, and cognitive archeology (as yet unpublished manuscript), Dept. of Psychology, University of Illinois at Urbana Champaign, 1985.

[Mic 83a] Michalski, R.S., A theory and methodology of inductive learning, *Artificial Intelligence 20*, 2 (1983), 111-161; reprinted in Michalski, R.S. et al (Ed.), *Machine Learning: An Artificial Intelligence Approach*, Tioga, 1983, 83-134.

[Mic 83b] Michalski, R.S. and Stepp, R.E., Learning from observation: Conceptual clustering, in Michalski, R.S. et al (Ed.), *Machine Learning: An Artificial Intelligence Approach*, Tioga, 1983, 331-363.

[Mit 83] Mitchell, T.M., Learning and problem solving, *Proc. Eighth International Joint Conference on Artificial Intelligence*, 1983, 1139-1151.

[Ni 80] Nilsson, N.J., *Principles of Artificial Intelligence*, Tioga, 1980.

[Re 76] Rendell, L.A., A method for automatic generation of heuristics for state-space problems, Dept of Computer Science CS-76-10, University of Waterloo, 1976.

[Re 77] Rendell, L.A., A locally optimal solution of the fifteen puzzle produced by an automatic evaluation function generator, Dept of Computer Science CS-77-36, University of Waterloo, 1977.

[Re 81] Rendell, L.A., An adaptive plan for state-space problems, Dept of Computer Science CS-81-13, (PhD thesis), University of Waterloo, 1981.

[Re 83a] Rendell, L.A., A new basis for state-space learning systems and a successful implementation, *Artificial Intelligence 20* (1983), 4, 369-392.

[Re 83b] Rendell, L.A., A learning system which accommodates feature interactions, *Proc. Eighth International Joint Conference on Artificial Intelligence*, 1983, 469-472.

[Re 83c] Rendell, L.A., A doubly layered, genetic penetrance learning system, *Proc. Third National Conference on Artificial Intelligence*, 1983, 343-347.

[Re 83d] Rendell, L.A., Conceptual knowledge acquisition in search, University of Guelph Report CIS-83-15, Dept. of Computing and Information Science, Guelph, Canada, 1983 (to appear in Bolc, L. (ed.), *Knowledge Based Learning Systems*, Springer-Verlag).

[Re 85a] Rendell, L.A., Utility patterns as criteria for efficient generalization learning, *Proc. 1985 Conference on Intelligent Systems and Machines*, (to appear), 1985.

[Re 85b] Rendell, L.A., A scientific approach to applied induction, Proc. 1985 International Machine Learning Workshop, Rutgers University (to appear), 1985.

[Re 85c] Rendell, L.A., Substantial constructive induction using layered information compression: Tractable feature formation in search, *Proc. Ninth International Joint Conference on Artificial Intelligence*, (to appear), 1985.

[Sa 63] Samuel, A.L., Some studies in machine learning using the game of checkers, in Feigenbaum, E.A. and Feldman, J. (Ed.), *Computers and Thought*, McGraw-Hill, 1963, 71-105.

[Sa 67] Samuel, A.L., Some studies in machine learning using the game of checkers II — recent progress, *IBM J. Res. and Develop. 11* (1967) 601-617.

[Sm 80] Smith, S.F., A learning system based on genetic adaptive algorithms, PhD Dissertation, University of Pittsburgh, 1980.

[Sm 83] Smith, S.F., Flexible learning of problem solving heuristics through adaptive search, *Proc. Eighth International Joint Conference on Artificial Intelligence*, 1983, 422-425.

APPENDIX A. GLOSSARY OF TERMS

Clustering. *Cluster analysis* has long been used as a tool for induction in statistics and pattern recognition [An 73]. (See "induction".) Improvements to basic clustering techniques generally use more than just the features of a datum ([An 73, p. 194] suggests "external criteria"). External criteria in [Mi 83, Re 76, Re 83a, Re 85b] involve prior specification of the forms clusters may take (this has been called "conceptual clustering" [Mi 83]). Criteria in [Re 76, Re 83a, Re 85b] are based on the data environment (see

"utility") below).[15] This paper uses clustering to create economical, compressed genetic structures *(genotypes)*.

Feature. A *feature* is an attribute or property of an object. Features are usually quite abstract (e.g. "center control" or "mobility") in a board game. The *utility* (see below) varies smoothly with a feature.

Genetic algorithm. In a GA, a the character of an *individual* of a *population* is called a *phenotype*. The phenotype is coded as a string of digits called the *genotype*. A single digit is a *gene*. Instead of searching rule space directly (compare "learning system"), a GA searches gene space (i.e. a GA searches for good genes in the population of genotypes). This search uses the merit μ of individual genotypes, selecting the more successful individuals to undergo genetic operations for the production of offspring. See §2 and Fig. 3.

Induction. *Induction or generalization learning* is an important means for knowledge acquisition. Information is actually *created*, as data are compressed into *classes* or *categories* in order to predict future events efficiently and effectively. Induction may create feature space neighborhoods or *clusters*. See "clustering" and §4.1.

Learning System. Buchanan et al. present a general model which distinguishes components of a learning system [Bu 78]. The *performance* element *PE* is guided by a *control structure* H. Based on observation of the PE, the *critic* assesses H, possibly localizing credit to parts of H [Bu 78, Di 81]. The *learning* element *LE* uses this information to improve H, for the next round of task performance. *Layered* systems have multiple PE's, critics, and LE's (e.g. PLS2 uses *PLS1* as its PE — see Fig. 1). Just as a PE searches for its goal in problem space, the LE searches in *rule space* [Di 82] for an optimal H to control the PE.

To facilitate this higher goal, PLS2 uses an intermediate knowledge structure which divides feature space into *regions* relating feature values to object *utility* [Re 83d] and discovering a useful subset of features (cf [Sa 63]). In this paper, the control structure H is a linear evaluation func-

---

15. A new learning system [Re 85c] introduces higher-dimensional clustering for creation of structure.

tion [Ni 80], and the "rules" are feature weights for H. Search for accurate regions replaces direct search of rule space; i.e. regions *mediate* data and H. As explained in §3, sets of regions become compressed GA "genotypes". See also "genetic algorithms", "PLS", "region", and Fig. 1.

Merit μ. Also called *payoff* or *fitness*, this is the measure used by a genetic algorithm to select parent genotypes for preferential reproduction of successful individuals. Compare "utility", also see "genetic algorithms".

Object. Objects are any data to be generalized into categories. Relationships usually depend on task domain. See "utility".

PLS. The *probabilistic learning system* can learn what are sometimes called "single concepts" [Di 82], but PLS is capable of much more difficult tasks, involving noise management, incremental learning, and normalization of biased data. PLS1 uniquely discovered locally optimal heuristics in search [Re 83a], and PLS2 is the effective and efficient extension examined in this paper. PLS manipulates "regions" (see below), using various inductive operations described in §4.

Region or Cell. Depending on one's viewpoint, the *region* is PLS's basic structure for clustering or for the genetic algorithm. The region is a compressed representation of a utility surface in *augmented* feature space; it is also a compressed genotype representing a utility function to be optimized. As explained in [Re 83d], the region representation is fully expressive, providing the features are. See §3 and Figs. 3 & 4.

Utility u. This is any measure of the usefulness of an object in the performance of some task. The *utility* provides a link between the task domain and PLS generalization algorithms. Utility can be a probability, as in Fig. 2. Compare *merit*. See §1, 3.

APPENDIX B. PLS1 LIMITATIONS

PLS1 alone is inherently limited. The problems relate to modification of the main knowledge structure, the cumulative region set R = {(r, u, e)}. As mentioned in §4.1, R undergoes two basic alterations. PLS1 gradually changes the *meaning* of an established feature space rectangle r by updating its associated utility u (along with u's error e). PLS1 also incrementally

refines the feature space, as rectangles r are continually split.

Both of these modifications (utility revision and region refinement) are largely directed by search data, but the degree to which newer information affects R depends on various choices of system parameters [Re 83a]. System parameters influence estimates of the error e, and determine the degree of region refinement. These, in turn, affect the relative importance of new versus established knowledge.

Consequently, values of these parameters influence task performance. For example, there is a tradeoff between utility revision and region refinement. If regions are refined too quickly, accuracy suffers (this is theoretically predictable). If, instead, utility revision predominates, regions become inert (their estimated errors decline), but sometimes incorrectly.[16]

There are several other problems, including difficulties in training, limitation in the utility revision algorithm, and inaccurate estimation of various errors. As a result, utility estimations are imperfect, and biased in unknown ways.

Together, the above uncertainties and sensitivities explain the failure of PLS1 always to locate an optimum with static training (Table I). The net effect is that PLS1 works fairly well with no parameter tuning and unsophisticated training, and close to optimal with mild tuning and informed training [Co 84], as long as the features are well behaved.

By nature, however, PLS1 requires features exhibiting no worse than mild interactions. This is a serious restriction, since feature nonlinearity is prevalent. On its own, then, PLS1 is inherently limited. There is simply *no way* to learn utility accurately unless the effects of differing heuristic functions H are compared, as in PLS2.

ACKNOWLEDGEMENTS

I would like to thank Dave Coles for his lasting enthusiasm during the development, implementation and testing of this system. I appreciate the helpful suggestions from Chris Matheus, Mike Mauldin, and the Conference Reviewers.

---

16. Although system parameters are given by domain-independent statistical analysis, tuning these parameters nevertheless improves performance in some cases. (This is not required in PLS2.)

Learning Multiclass Pattern Discrimination

J. David Schaffer
Department of Electrical Engineering
Vanderbilt University
Nashville, TN 37235

ABSTRACT

Genetic algorithms (GA's) are powerful, general purpose adaptive search techniques which have been use successfully in a variety of learning systems. Previous implementations have tended to use scalar feedback concerning the performance of alternate knowledge structures on the task to be learned. This approach was found to be inadequate when the task was multiclass pattern discrimination. By providing the GA with multidimensional feedback, a problem of this type was successfully learned. In addition, a careful balance of reward and punishment was found to be necessary in order to guide the opportunistic GA to a correct solution of the problem.

1. Introduction

This paper presents some results of experiments with a software system designed to learn rules for multiclass pattern discrimination from examples of correctly classified patterns.

The original motivation for this research arose from attempts to develop computer programs capable of intelligent signal analysis. One such application domain is computer analysis of bioelectric signals such as EMG's and EEG's. Previous attempts to model the actions of an electroencephalographer using variations of traditional electrical engineering approaches had met with some success, but complete agreement with the human expert eluded us [2,4]. Attempts to elicit the knowledge from the expert for use by an expert system had similarly met with limited success [5]. Nevertheless, it was clear that the expert was able to reliably preform this complex pattern discrimination, even if he was unable to completely articulate how he did it. Therefore, an algorithm capable of inferring rules for discrimination from examples of correctly classified patterns, seemed to hold promise.

A search of the literature for methods by which machines could learn rules from examples revealed a small number of currently active approaches [11,12]. Of these, the methods based on Holland's Genetic Adaptive Plans [9], or Genetic Algorithms (GA's) seemed to hold the most promise for the following reasons. They have been shown both theoretically [9] and empirically [6,7,8] to take near optimal advantage of the information gained during attempts to solve a problem. In addition, the preferred coding for the example patterns is as bit strings. This offers the possibility that one may avoid the usual feature extraction processes, which, although capable of considerable data reduction, also carry with them the risk that the reduced feature set may no longer contain the information contained in the original signal which made the discrimination possible. A GA might be developed that operates on the raw digitized signals.

For the remainder of this paper, an understanding of the basics of GS's has been assumed. They are well described elsewhere [3,6,7,8,9,10,13].

2. Background

There are two learning systems based on GA's in the literature which might be considered immediately ancestral to the research described herein. The Cognitive System One (CS-1) of Holland and Rietman [10] was the first published account of a system which combined the computational power of a production system (PS)with a GA-based learning component. This system exhibited an ability to learn a dual-reward linear maze. The Learning System One (LS-1) of Smith [13] took this concept further and demonstrated learning behavior in two different problem domains, the maze problem and draw poker playing.

LS-1 appeared to have an important advantage over CS-1. CS-1 maintained a population of knowledge structures which were individual PS rules, which gave rise to the credit assignment problem. A heuristic method had to be devised to distribute credit for rewarded behavior among the rules which cooperated in producing that behavior. LS-1 avoided this problem by using complete rule sets, or PS programs, as the individuals in its populations. A difficulty with this approach involved the use of scalar evaluations for the individuals. When the task to be learned is multidimensional, then scalarization of the feedback to tha GA creates difficulties which will be described below.

3. A Vector Extension of LS-1 (LS-2)

Several attempts to learn a multiclass pattern discrimination problem with LS-1 were unsuccessful. Examination of the populations of programs in both the early and late stages of the searches revealed a common pattern. Knowledge of how to recognize a particular class was frequently absent from the populations in the latter stages of the search even when such knowledge was present in earlier populations. The hypothesis for this was that the scalar feedback was forcing competition between programs whose knowledge was complementary.

Consider this simple example. Suppose the feedback to the GA consists essentially of the number of training cases correctly classified. Suppose program P1 contains rules which correctly classify classes A and B while program P2 classifies only instances in class C. If all classes are equally represented in the training set, then P1 will appear to be twice as "fit" as P2. In a survival-of-the-fittest selection process, specialized knowledge such as that possessed by P2 may die out.

The solution to this problem was to modify the critic so that a vector of performances could be reported back to the GA, one for each facet of the problem (class of pattern to be learned, in this case). The selection process of the GA was also modified in such a way that an independent survival-of-the-fittest selection is performed on each dimension of the performance vector, each time selecting only the appropriate fraction of the population. This selection is performed with replacement so that individuals with better that average performance on more than one dimension have the appropriate probability of multiple selection while simultaneously protecting individuals with specialized knowledge from unfair competition. This represents a simple generalization of the traditional selection process which reverts to the traditional process when the number of dimensions of performance is one.

4. The LS-2 Production System

4.1 Knowledge Representation

The binary coding scheme for the IF-THEN rules was devised by Smith and called Knowledge Structure One (KS-1). On the IF side of each rule a number of clauses which are sensitive to external signals are allowed as well as a number of clauses sensitive to internal signals i.e. signals from other rules. These clauses are simply strings on the alphabet {0,1,#} where 0 and 1 require an exact match and # will match either 0 or 1. For example, the clause 1## would match any 3-bit signal whose first bit is a 1. On the THEN side, each rule has a signal which it deposits in short term memory if the rule fires. From there, it becomes accessible to other rules. Also on the THEN side is an action which is performed if the rule fires and also survives the conflict resolution at the end of the production system cycle. This general scheme as well as an example of the binary coding of these rules is shown in figure 1.

An individual for the purposes of the GA is a set of rules in KS-1 format concatenated together. For pattern discrimination problems, each rule would have one detector (external signal) clause and one message (internal signal) clause. The length of the detector clause is set to the length of the patterns to be learned and the length of the message clause would be set to the length of the messages. The message length need only be a function of the maximum number of rules an individual may contain, the only requirement being that there should be enough bits so that every rule

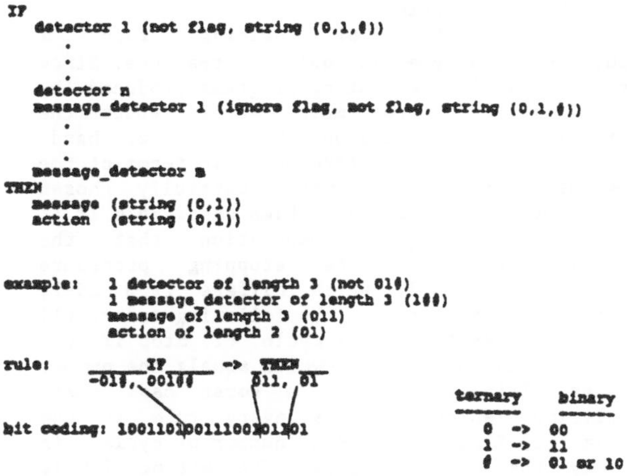

Figure 1. KS-1 Knowledge Representation Scheme

might have a unique signal. The actions would include one for each class of pattern to be discriminated with one for no-operation (noop). The noop action is a simple device which allows for the evolution of sets of rules which perform a chain calculation leading to a final classification decision.

It may be noticed that the binary coding scheme illustrated in figure 1 admits a certain redundancy. By having two binary codings for the don't care symbol (# = 01 or 10), there are many binary codings possible for any given rule, with the problem growing worse as the number of #'s increases. This observation gave rise to the conjecture that the knowledge structures might be coded directly in their natural ternary (0,1,#) alphabet. To counter this conjecture, is the known superiority of binary coding for the gathering of information from the hyperplanes of the search space (see Holland [10,p71] or Smith [13,p56]). However, it was unclear how these two arguments, might be compared and the better coding selected. Therefore, several tests were conducted wherein the same problems were solved by LS-2 using first binary coding and then ternary coding.

4.2 Recognize-act Cycle

The production system contained the usual recognize-act cycle with parallel firing of rules allowed in the following fashion. On each cycle, every rule is tested for a match of its IF-side clauses. Every rule which does match, "fires" in the sense that its message is posted in short term memory, but its action is merely tallied in an array called the suggested-action array. Only at the end of the cycle is an action selected and actually performed. If the number of "real" (non noop) actions suggested is zero, then another cycle is initiated. If one action is suggested, then it is performed. If more than one action is suggested, then a stochastic conflict resolution scheme is invoked which randomly selects one of the suggested actions with the probability of selection being proportional to the number of rules suggesting it.

4.3 The Halting Problem

The question of halting such a computational scheme is quite a real one. Since this system will be executing programs produced by genetic search, one must worry about the possibility of infinite loops. On the other hand, an arbitrary stopping threshold in terms of the number of cycles to allow must be carefully chosen so as not to render a problem unsolvable if it requires more lengthy computation that the threshold permits. The stopping procedure implemented for LS-2, which differs from LS-1, examines four criteria in the following order: (1) Stop if no rules fired this cycle. (2) Stop if the number of consecutive noop cycles equals the number of rules. This allows for a worst case chain calculation which utilizes every rule in the program. (3) Stop if the total number of cycles is N times the number of rules. The setting of N is the arbitrary threshold just mentioned and is currently set to 3, but at least the actual stopping threshold is also a function of the number of rules. (4) Stop if the task has been learned.

5. The Critic

Besides vector performance and the stopping criteria, the other major area of difference between LS-1 and LS-2 was that of the critic. Smith anticipated a problem with a critic which only rewarded task successes. In the early stages of a genetic search, especially one started with a random population, successes are likely to be rare and thus such a critic would be unable to give the GA any information about which population members were more promising than others. In the absence of such information the GA would revert to near random search. In addition, in the later stages of the search when successes were plentiful, some way of identifying the better (e.g. more parsimonious) individuals would lead more efficiently to good solutions. Smith sought measures to provide this kind of information which were also task independent. He devised two classes, static measures which could be computed just by examining the rule set, and dynamic measures which could be computed only by monitoring the action of the rule set on the task. The static measures included such items as the amount of interrule communication measured by how many rule messages would match the message clause of other rules, the generality of the clauses measured by the number of # symbols and so on. The dynamic measures included the amount of random behavior measured by the activity of the conflict resolution procedure and the percentage of inactive rules. While these measures have an intuitive appeal, it is not clear that they are always associated with superior performance regardless of the task. In addition, some preliminary experiments with LS-2 using these measures as an independent dimension of the performance vector revealed a poor correlation with task success. These measures, then, were dropped from LS-2 in favor of the critic described below. This critic did incorporate the amount of random activity which was called guessing behavior.

The properties sought for a critic for pattern discrimination learning were similar to those identified by Smith. In the early stages of

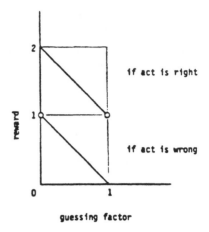

Figure 2. The Reward Function for the First Critic

the search, guessing should be encouraged so as to locate as quickly as possible, some knowledge which seems to work. However, the same critic must discourage guessing in the later stages when more reliable classification becomes the goal. The first critic reward function designed with these properties is illustrated in figure 2. The guessing factor is a measure of the uncertainty with which the production system reaches a conclusion as measured by the number of rules suggesting some other conclusion. This reward function would give some credit for a wrong conclusion if the right action were at least suggested and would also give less than maximum credit for the right action if it were only guessed. While this critic seemed reasonable, it had a subtle weakness which was revealed by experimentation. Better critics were later designed, but they will be discussed with the results which lead to them.

6. The Test Problems

6.1 The Parity Problem

As a direct test that the GA is capable of producing more powerful programs than perceptrons, the parity problem was included. This problem involves the discrimination of two classes which are inextricably mixed in the feature space as shown in figure 3. Although a discriminator requiring linear separability in the feature space would be unable to solve this problem, a system able to compute the parity (a derived feature) of the given features would be able to solve this problem easily.

6.2 A Multiclass Pattern Problem

As a test of a GA on a multiclass signal discrimination learning task a small problem was selected from the literature which used EMG signals. Bekey at al [1] measured the neural firing patterns enervating six muscles in the lower leg and then reduced the signal from each muscle to a two-bit string indicating that the muscle was either on (1) or off (0) during two phases of the gait cycle. Thus, each subject tested yielded a

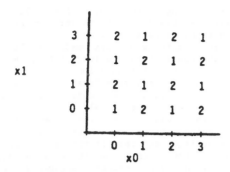

Figure 3. The Parity Problem

12-bit string which was classified by the clinician as belonging to one of five classes of gait as shown in figure 4.

The 11 patterns shown in figure 4 were considered by Bekey and his colleagues to be typical of their classes and were used as a training set for a statistical discrimination experiment. This training set has several features to recommend it as a test bed for a GA learning experiment. The strings were short and binary coded. Also the training set is small implying a small computational burden. Although the problem is underconstrained (there are many possible solutions), it is rich enough to frustrate a simple linear discriminator like a perceptron.

7. Results

7.1 The Parity Problem

After trying 5040 different PS programs (evaluations)on the parity problem, LS-2 found the solution shown in figure 5. The training set for this problem consisted of all 16 patterns shown in figure 3 with x0 and x1 coded at two bits each. The "ignore" in figure 5 indicates that the message clauses in all of the rules were turned off. Thus, the solution did not require any interrule communication. Instead two rules were evolved for each combination of low order bits in the feature strings. The 2-bit strings to the right of the -> symbol in figure 5 are the signals placed in short term memory which are ignored by the other rules. The rightmost number in each rule is the action (classification asserted for the pattern if the rule fires).

7.2 The EMG Problem

The first trials with the EMG problem clearly indicated the superiority of the KS-1 binary coding over the ternary coding. Comparison runs to discriminate different pairs of classes and one run to discriminate a 3-class subset of the Bekey training set all learned the discrimination faster with binary coding than with ternary coding. Besides this series, all other experiments reported in this paper utilized binary coding.

A solution to a 2-class subproblem of the Bekey set was located in 2520 evaluations. However, a 3-class subproblem required almost 30000 evaluations and the full 5-class problem remained unsolved after 100000 evaluations. This rapid increase in computational effort soon drew attention to the critic as a possible cause. By examining some of the individuals in different populations, the survival strategy that LS-2 was adopting was discovered. It was observed that very early in the searches, even in the initial populations, individuals appeared who received maximum credit for one class of patterns and zero for all other classes. These were named specialists since they seemed to have knowledge of how to identify one pattern class without guessing, but no others. However, close examination revealed their simple strategy. Each specialist contained a single over-general rule which fired for every pattern, calling class x. If the pattern was class x, the critic assigned maximum credit, if not then the critic assigned zero. However, under the vector survival-of-the-fittest, maximum fitness on one dimension was sufficient to assure survival. These specialists also tended to persist since their genes combined poorly, not producing offspring able to perform well on more than one dimension. This strategy of wild guessing was encouraged by the critic; some form of punishment for wrong actions was clearly needed.

The original critic was then replaced by one which simply assigned two points for each correct suggestion and minus one for each wrong suggestion. More credit for right than wrong seemed appropriate in light of the need to preserve individuals with some good knowledge, even if imperfect. This critic also embodied a shift of focus from the overt actions of the individual to the suggested actions, thus rewarding the "thinking" more directly.

Under the new critic, LS-2 solved the 2-class problem in only 1440 evaluations, a 43% reduction in search effort. However, this advantage did not hold for higher order problems. Again, an examination of the individuals persisting in the population revealed the new survival strategy. With a 2-class task, the expectation for wild guessing under this critic, is one half the maximum credit so long as each class in equally represented in the training set. However, as the number of classes increases, this expectation decreases and even becomes negative for fairly modest tasks. When

```
class 1 (normal)
    (0,1,1,1,1,1,1,0,0,0,0,0)
class 2 (equinus)
    (0,1,1,1,1,1,0,0,0,0,0,0)
    (0,1,1,0,1,1,1,0,0,0,0,0)
    (0,1,1,1,0,1,1,1,0,0,0,0)
    (0,1,1,1,0,1,0,0,0,1,0,0)
class 3 (flat footed)
    (0,0,0,0,1,1,1,0,0,0,0,0)
    (0,1,0,0,0,1,1,1,0,0,0,1)
class 4 (varus)
    (1,0,1,1,0,0,1,0,0,1,0,0)
    (1,1,1,0,1,1,0,0,0,0,0,0)
class 5 (valgus)
    (0,0,1,0,1,1,1,0,0,0,1,1)
    (0,1,1,0,0,1,1,0,0,0,0,1)
```

Figure 4. The Bekey Training Cases

	IF		-> THEN
rule 0:	#1#1	ignore ->	01 1
rule 1:	#0#1	ignore ->	01 2
rule 2:	#0#0	ignore ->	10 1
rule 3:	#1#0	ignore ->	10 2

Figure 5. The Solution to the Parity Problem

this occurs, LS-2 quickly begins evolving individuals with no rules that fire at all. Doing nothing at least scores zero which is better than being punished. A balance of reward and punishment which will be maintained as tasks increase in complexity is needed so as to avoid the GA's ability to quickly exploit this weakness in the critic function.

The next critic employed a computational scheme based on that used on the Scholastic Apptitude Test and so was called SAT scoring. The main idea in the scoring of multiple choice tests is that indiscriminant guessing should have an expectation of zero, but that if a student can eliminate some of the choices on a question, then he should be encouraged to guess by having the expected score increase as the range of guessing decreases. For the SAT, this is achieved by subtracting from the number of correct answers, the number of wrong answers weighted by the inverse of the number of choices minus one. This gives an expectation which varies from zero for wild guessing to the maximum score for no guessing. For LS-2 a slightly different expectation was thought desirable. Wild guessing was deemed better than doing nothing because this at least would give the GA some active rules to deal with. So the designed expectation was that wild guessing (e.g. calling every case the same class) should score half of the maximum.

At this point in the experimentation, an effort was also initiated to learn about the sensitivity of LS-2 to changes in four of its main parameters, population size, crossover, mutation and inversion rates. All experiments reported so far used a population size of 30 per dimension of the performance vector, a crossover rate of .95, a mutation rate of .01 and an inversion rate of .25. These first three values were suggested by Greffenstette [8] and the inversion rate by Smith [13]. Limited resources prevented the best approach which would have been the meta-GA approach of Greffenstette, so different settings were produced by increasing the population size in steps of 10 per dimension and simultaneously reducing the rates more or less in unison. This process was continued until the mean evaluations-to-solution stopped improving. Means were computed for three runs at each setting with different random seeds.

The SAT critic has the same expectation as the previous critic for 2-class problems with balanced training sets, so this task was not repeated. A 3-class subproblem was solved in 6921 evaluations, a 77% improvement over the original critic. A 4-class subproblem was solved in 26591 evaluations. Both of these results represented a

best parameter setting of 40, .90, .005 and .20 for population size per dimension, crossover, mutation and inversion rates respectively.

One final improvement was made in LS-2, this time to the conflict resolution. In LS-1 Smith had not permitted conflict resolution to consider the noop action so long as a "real" action were suggested. In all the LS-2 experiments so far, noop competed equally with the "real" actions. The argument for this was that for some task environments, doing nothing, or continuing to think (cycle) was a decision and that if the environment were dynamic, then this might well affect performance. However, some counter arguments can also be made. The pattern discrimination tasks considered so far are not dynamic; the patterns don't change while LS-2 is trying to decide. Also, this strategy allows for some stochastic effect to remain in the critic-reported values. By deciding to cycle again when a "real" action had been suggested, LS-2 postponed the computation of the credit in a non-deterministic way. The critic was only permitted to evaluate the suggested action array on the final cycle. I would now argue that if the task environment is dynamic, and a do nothing action should be considered, then it should be explicitly included as one of the "real" actions. Noop should not be considered a do nothing action.

With this final improvement, LS-2 solved the 3-, 4- and full 5-class problems in 5647, 15938 and 44509 evaluations respectively. The taming effect of these improvements in LS-2 are illustrated in figure 6.

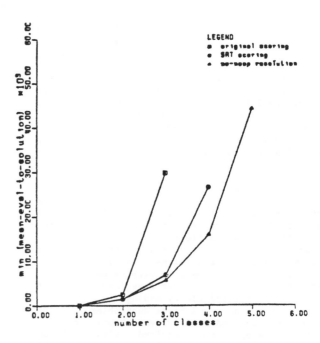

Figure 6. Improvements in LS-2 with Changing Critic

78

8. Discussion

The major finding of this research was that vector feedback is essential to multiclass discriminant learning. Vector selection provides the necessary protection against unfair competition while simultaneously providing the proper pressure for the evolution of the utopian individual capable of high performance on all facets of the task.

Secondary to this major finding are a number of observations which may contribute to better understanding of GA's and how to effectively utilize them.

The solution of the parity problem clearly demonstrates LS-2's ability to learn non-linear discrimination.

Ternary coding of KS-1 was inferior to binary coding, even with the redundancy inherent in the binary coding scheme. A search for coding schemes which are binary and yet avoid this redundancy might pay handsome dividends.

Grefenstette's finding [8] that genetic search may be very efficient with smaller populations and higher mixing rates than previous wisdom suggested, seems generally to have been confirmed. Populations of 40 per dimension of performance with crossover rates of .7 to .9, mutation rates of .001 to .01 and inversion rates of .1 to .2 provided the best performance on the problems studied here. It should be noted, however, that the search was limited and began with Grefenstette's solution.

As Smith observed, the critic is critical. The GA is capable of exploiting the properties of its critic and so good performance was only achieved when reward and punishment were carefully balanced. The application of punishment to a performance vector has raised a question which did not occur with scalar performance systems. There are two places where this punishment may be applied. Suppose that a PS program incorrectly classifies a class 1 case as class 2. By applying the punishment to the class 1 slot of the performance vector one is punishing the failure to do the right thing. By applying it to the class 2 slot, one is punishing the program for doing the wrong thing. It is unknown which strategy, or both leads to faster learning. The experiments reported here applied the punishment to the slot corresponding to the case to be classified, thus always punishing the failure to do the right thing. Other approaches might be profitably studied.

The task independent measures proposed by Smith did not seem to be sufficiently closely associated with good performance to warrant their use. However, his strategy of disallowing noop actions to compete in conflict resolution was superior to allowing it.

A final observation is in order on the original question of using a GA for intelligent signal classification. The strategy used in LS-2 seems to be promising, but requires that a prior decision be made on the length and sampling rate for the signal. The patterns must be "frozen" so that the system can examine them. This feature seems to impose undesirable limitations. A more dynamic method of examining the signal, bit by bit, and only reporting a decision when enough information has been acquired to do so with confidence, seems to offer a more robust approach.

REFERENCES

1. A.B. Bekey, C. Chang, J. Perry, and M.M. Hoffer, "Pattern recognition of multiple EMG signals applied to the description of human gait," Proceedings of the IEEE, Vol. 65 No. 5, May 1977.

2. J.R. Bourne, V. Jagannathan, B. Hamel, B.H. Jansen, J.W. Ward, J.R. Hughes and C.W. Erwin "Evaluation of a syntactic pattern recognition approach to quantitative electroencephalographic analysis," Electroencephapography & Clinical Neurophysiology, 52:57-64, 1981.

3. A. Brindle, Genetic Algorithms for function optimization, Ph.D. Dissertation, University of Alberta, Edmonton, Alberta, Canada, 1975.

4. D.A. Giese, J.R. Bourne and J.W. Ward, "Syntactic analysis of the electroencephalogram," IEEE Trans. Systems, Man and Cybernetics, Vol. SMC-9 No. 8, Aug 1979.

5. V. Jagannathan, An artificial intelligence approach to computerized electroencephalogram analysis, Ph.D. Dissertation, Vanderbilt University, Nashville, Tennessee 1981.

6. Kenneth DeJong, Analysis of the behavior of a class of genetic adaptive systems, Ph.D. Dissertation, University of Michigan, Ann Arbor, 1975.

7. Kenneth DeJong, "Adaptive system design: a genetic approach," IEEE Trans. Systems, Man and Cybernetics, Vol. SMC-10 No. 9, Sept 1980.

8. John J. Greffenstette, "Genetic algorithms for multilevel adaptive systems," IEEE Trans. on Systems, Man and Cybernetics, in press.

9. John H. Holland, Adaptation in natural and artificial systems, University of Michigan Press, Ann Arbor, Michigan 1975.

10. J.H. Holland and J.S. Reitman, "Cognitive systems based on adaptive algorithms," in Pattern-Directed Inference Systems, Waterman and Hayes-Roth (Eds.), Academic Press, 1978.

11. R.S. Michalski, J.G. Carbonell and T.M. Mitchell, (Eds.), Machine Learning, Tioga Publishing Co., Palo Alto, California 1983.

12. R.S. Michalski, J.G. Carbonell and T.M. Mitchell, (Eds.), Proceedings of the International Machine Learning Workshop, University of Illinois, Urbana-Champaign, Illinois, 1983.

13. S.F. Smith, A learning system based on genetic adaptive algorithms, Ph.D. Dissertation, University of Pittsburg, 1980.

IMPROVING THE PERFORMANCE OF GENETIC ALGORITHMS
IN CLASSIFIER SYSTEMS

Lashon B. Booker

Navy Center for Applied Research in AI
Naval Research Laboratory, Code 7510
Washington, D.C. 20375

ABSTRACT

Classifier systems must continuously infer useful categories and other generalizations — in the form of classifier taxa — from the steady stream of messages received and transmitted. This paper describes ways to use the genetic algorithm more effectively in discovering such patterns. Two issues are addressed. First, a flexible criterion is advocated for deciding when a message matches a classifier taxon. This is shown to improve performance over a wide range of categorization problems. Second, a restricted mating policy and crowding algorithm are introduced. These modifications lead to the growth and dynamic management of subpopulations correlated with the various pattern categories in the environment.

INTRODUCTION

A *classifier system* is a special kind of production system designed to permit non-trivial modifications and reorganizations of its rules as it performs a task [Holland,1976]. Classifier systems process binary messages. Each rule or *classifier* is a fixed length string whose activating condition, called a *taxon*, is a string in the alphabet $\{0,1,\#\}$. The differences between classifier systems and more conventional production systems are discussed by Booker [1982] and Holland [1983].

One of the most important qualities of classifiers systems as a computational paradigm is their flexibility under changing environmental conditions [Holland,1983]. This is the major reason why these systems are being applied to dynamic, real-world problems like the control of combat systems [Kuchinski,1985] and gas pipelines [Goldberg,1983]. Conventional rule-based systems are brittle in the sense that they function poorly, if at all, when the domain or underlying model changes slightly. Several factors work together to enable classifier systems to avoid this kind of brittleness: parallelism, categorization, active competition of alternative hypotheses, system elements constructed from "building blocks" , etc.

Perhaps the most important factor is the direct and computationally efficient implementation of categorization. Holland [1983, p.92] points out that

> Categorization is the system's *sine qua non* for combating the environment's perceptual novelty.

Classifier systems must continuously infer useful categories and other generalizations — in the form of taxa — from the steady stream of messages received and transmitted. This approach to pattern-directed inference poses several difficulties. For example, the number of categories needed to function in a task environment is usually not known in advance. The system must therefore dynamically manage its limited classifier memory so that, as a whole, it accounts for all the important pattern classes. Moreover, since the categories created depend on which messages are compared, the system must also determine which messages should be clustered into a category.

The fundamental inference procedure for addressing these issues is the genetic algorithm [Holland,1975]. While genetic algorithms have been analyzed and empirically tested for years [DeJong,1975;Bethke,1981], most of the knowledge about how to implement them has come from applications in function optimization. There has been little work done to determine the best implementation for the problems faced by a classifier system. This paper begins to formulate such an understanding with respect to categorization. In particular, two questions related to genetic algorithms and classifiers systems are examined:

(1) What kinds of performance measures provide the most informative ranking of classifier taxa, allowing the genetic algorithm to efficiently discover useful patterns?

(2) How can a population of classifier taxa be dynamically partitioned into distinguishable, specialized subpopulations correlated with the set of categories in the message environment?

Finding answers to these and related questions is an important step toward improving the categorization abilities of classifier systems; and, expanding the repertoire of problems these systems can be used to solve.

THE CATEGORIZATION PROBLEM

In order to formulate these issues more precisely, we begin by specifying a class of categorization problems. Subsequently, a criterion is given for evaluating various solutions to one of these problems.

Defining Message Categories

Hayes-Roth [1973] defines a "schematic" approach to characterizing pattern categories that has proven useful in building test-bed environments for classifier systems [Booker,1982]. This approach assumes, in the simplest case, that each pattern category can be defined by a single structural prototype or *characteristic*. Each such characteristic is a schema designating a set of features values required for category membership. Unspecified values are assumed to be irrelevant for determining membership.

The obvious generalization of using just one characteristic to define a category is to permit several characteristics to define a category disjunctively. Pattern generators based on the schematic approach generate exemplars by assigning the mandatory combinations given by one or more of the pattern characteristics and producing irrelevant feature values probabilistically. In this way, each exemplar of a category manifests at least one of the defining characteristics. The categorization problem can be very difficult under the schematic approach since any given item can instantiate the characteristics of several alternative categories.

Classifiers receive, process, and transmit binary message strings. We define a category of binary strings by specifying a set of pattern characteristics. Each characteristic is a string in the alphabet $\{1,0,*\}$ where the $*$ is a place holder for irrelevant features. A characteristic is a template for generating binary strings in the sense that the 1 and 0 indicate mandatory values and the $*$ indicates values to be generated at random. Thus the characteristic $1*0*$ generates the four strings 1000, 1001, 1100, and 1101. When more than one characteristic is associated with a category, one is selected at random to generate an exemplar. The correspondence between the syntax of a taxon and the designation of pattern characteristics is obvious. The class of pattern categories defined in this manner therefore spans the full range of categorization problems solvable with a set of taxa.

An Evaluation Criterion

A set of taxa is a solution to a categorization problem if it corresponds directly with the set of characteristics defining the category. In this sense, the set of taxa models the structure of the category. One way to evaluate how closely a set of taxa models a set of characteristics is to define what an "ideal" model would look like; then, measure the discrepancy between the model given by the set of taxa and that ideal.

More specifically, the structure of a pattern category is given by its set of characteristics. We first consider the case involving only one characteristic. As the genetic algorithm searches the space of taxa, the collection of alleles and schemata in the population become increasingly less diverse. Eventually, the best schema and its associated alleles will dominate the population in the sense that alternatives will be present only in proportions roughly determined by the mutation rate. A population with this property will be called a perfect model of the category. The taxon which corresponds exactly with the characteristic will be called the perfect taxon.

One way to describe the perfect model quantitatively is in terms of the probability of occurrence for the perfect taxon. An exact value for this probability is difficult to compute, but for our purposes it can be approximated by the "steady state" probability[1] $P(\xi) = \prod_j P(\xi_j)$, where $P(\xi_j)$ is the proportion of the allele occurring at the jth position of the perfect taxon ξ. In the ideal case, if μ is the mutation rate, what we want is $P(\xi_j) = 1-\mu$ for the alleles of ξ. In order to measure the discrepancy between an arbitrary population and the perfect model, we can use the following metric:

$$G = P(\xi) \ln\left[\frac{P(\xi)}{P'(\xi)}\right] + (1-P(\xi)) \ln\left[\frac{(1-P(\xi))}{(1-P'(\xi))}\right]$$

where $P(\xi)$ is the ideal probability of occurrence for ξ and $P'(\xi)$ is ξ's probability of occurrence in the current population. This information-theoretic measure is called the *directed divergence* between the two probability distributions [Kullback,1959]. It is a non-negative quantity that approaches zero as the "resemblance" between P and P' increases. The G metric has proven useful in evaluating other systems that generate stochastic models of their environment (eg. Hinton *et al.* [1984]).

When a pattern category is defined by more than one characteristic, we can use the G metric to evaluate the population's model of each characteristic separately. This involves identifying the subset of the population involved in modeling each characteristic; and, treating each subset as a separate entity for the purpose of making measurements. A method for identifying these subsets will be discussed shortly.

MEASURES FOR RANKING TAXA

Given a class of categorization problems to be solved, and a criterion for evaluating solutions, we are now ready to examine the performance of the genetic algorithm. The starting point will be the measures used to rank taxa. Only if the taxa are usefully ranked can the genetic algorithm, or any learning heuristic, have hope of inferring the best taxon. In this section we first point out some deficiencies in the most often used measure; then, alternative measures are considered and shown to provide significantly better performance.

Brittleness and Match Scores

The first step in the execution cycle of every classifier system is a determination of

---

[1] The probability of occurrence under repeated crossover with uniform random pairing, in the absence of other operators.

which classifiers are relevant to the current set of messages. Most implementations make this determination using the straightforward matching criterion first proposed by Holland and Reitman [1978]. More specifically, if $M = m_1 m_2 \cdots m_k$, $m_i \in \{0,1\}$ is a message and $C = c_1 c_2 \cdots c_k$, $c_i \in \{0,1,\#\}$ is a classifier taxon, then the message M satisfies or *matches* C if and only if $m_i = c_i$ wherever c_i is 0 or 1. When $c_i = \#$, the value of m_i does not matter. Every classifier matched by a message is deemed relevant. Relevant classifiers are ranked according to the specificity of their taxa, where specificity is proportional to the number of non-#'s in the taxon. Holland and Reitman used a simple *match score* to measure relevance. The score is zero if the message does not match the taxon; otherwise it is equal to the number of non-# positions in the taxon.

This simple match score — hereafter called M1 — effectively guides the genetic algorithm in its search of relevant taxa. Because all non-relevant taxa are assigned a score of zero, however, M1 is the source of a subtle kind of brittleness. Whenever a message matches no taxon in the population, the choice of which taxa are relevant must be made at random. This can clearly have undesirable consequences for the performance of the classifier system; and, also for the prospects of quickly categorizing that message using the genetic algorithm.

In order to circumvent this difficulty, Holland and Reitman use an initial population of classifiers having a 90% proportion of #'s at each taxon position. This makes it very likely that relevant taxa will be available for the genetic algorithm to work with. Unless the pattern categories in the environment are very broad, though, the brittleness of this approach is still a concern. Suppose, for example, a classifier system must categorize the pattern characteristic 11010**. A fairly well-adapted population of

classifiers will contain taxa such as 11010##, 1#010##, 11#10#1, 11#10#0, etc. As the categorization process under the genetic algorithm continues, the variability in the population decreases. It therefore becomes unlikely that the population will contain many taxa having four or more #'s. Such taxa would have a match score too low to compete over the long run and survive. Now suppose the environment changes slightly so that the characteristic is **010**; that is, the category has been expanded to allow either a 0 or 1 in the first two positions. In order to consistently match the exemplars of the new category, the population needs a taxon with four #'s at exactly the right loci. There is no reason to expect such good fortune since the *combinations* of attribute values are no longer random. The population will most likely have no taxon to match new exemplars, and the genetic algorithm will blindly search for a solution.

Another proposed resolution of this dilemma is to simply insert the troublesome message into the population as a taxon [Holland,1976], perhaps with a few #'s added to it. The problem with this is that the rest of the classifier must be chosen more or less at random. By abandoning the "building block" approach to generating classifiers, this method introduces the brittleness inherent in ad hoc constructions that cannot make use of previous experience. What is needed is a way of determining *partial* relevance, so the genetic algorithm can discover useful building blocks even in taxa that are not matched. In the example cited above, such a capability would allow the genetic algorithm to recognize #1010## and 1#010## as "near miss" categorizations and work from there rapidly toward the solution ##010##.

Alternatives to M1

The brittleness associated with the match score M1 has a noticeable impact on categorization in classifier systems. To demonstrate this effect, a basic genetic algorithm [Booker,1982] was implemented to manipulate populations of classifier taxa. Taxa in this system are 16 positions long. The effectiveness of a match score in identifying useful building blocks is tested by presenting the genetic algorithm with a categorization problem. Each generation, a binary string belonging to the category is constructed and match scores are computed for every taxon. The genetic algorithm then generates a new population, using the match score to rate individual taxa.

To test M1, three pattern categories were selected:

C1 = 1111111111111111
C2 = 11111111********
C3 = 1***************

These characteristics are representative of the kinds of structural properties that are used to define categories, from the very specific to the very broad. Three sets of tests were run, each set starting with an initial population containing a different proportion of #'s. Each test involved a population of size 50 observed for 120 generations, giving a total of 6000 match score computations.[2] At the end of each run, a *G* value was computed for the final population to evaluate how well the characteristic had been modeled. The results of these experiments — averaged over 15 runs — are given in Table 1. For each pattern category, there are statistically significant[3] decreases in performance as the proportion of #'s is changed from 80% to 33% (Recall that the best *G* value is zero). Given this quantitative evidence of M1's brittleness, it is reasonable to ask if there are better performing alternatives.

The primary criterion for an alternative to M1 is that it identify useful building

Table 1 Final Average **G** Value Using M1			
Category	Initial Percentage of #'s		
	80%	50%	33%
C1	7.83	10.28	12.25
C2	4.95	16.72	25.13
C3	5.98	13.67	36.57

blocks in non-matching taxa; and, that it retain the strong selective pressure induced by M1 among matching taxa. One way to achieve this is to design a score that is equal to M1 for matching taxa, but assigns non-matching taxa values between 0 and 1. The question is, how should the non-matching taxa be ranked?

If we are concerned with directly identifying useful alleles, the following simple point system will suffice: award 1 point for each matched 0 or 1, ¾ point for each #, and nothing for each position not matched. The value for # is chosen to make sure it is more valuable for matching a random bit in a message than a 0 or 1, whose expected value in that case would be ½. To convert this point total into a value between 0 and 1, we divide by the square of the taxon length. This insures that there is an order of magnitude difference between the lowest score for a matching taxon and all scores for non-matching taxa. More formally, if l is the length of a taxon, n_1 is the number of exactly matched 0's and 1's, and n_2 is the number of #'s, we define a new match score

$$M2 = \begin{cases} M1 & \textit{if the message matches the taxon} \\ \dfrac{n_1 + ¾n_2}{l^2} & \textit{otherwise} \end{cases}$$

Another way to rank non-matching taxa is by counting the number of

---

[2] 6000 function evaluations is the observation interval used by DeJong [1975] that has become a standard in analysing genetic algorithms.

[3] For all results presented in this paper, this means a *t* test was performed comparing the means of two groups. The

alpha level for each test was .05.

mismatched 0's and 1's. This approach measures the Hamming distance between a message and a taxon for the non-# positions. A simple match score M3 can be defined to implement this idea. If n is the number of mismatched 0's and 1's, then

$$M3 = \begin{cases} M1 & \text{if the message matches the taxon} \\ \dfrac{l-n}{l^2} & \text{otherwise} \end{cases}$$

Now it must be determined if M2 and M3 usefully rank non-matching taxa; and, if so, whether that gives them an advantage over M1. Accordingly, M2 and M3 were tested on the same three patterns and types of populations described above for M1. These experiments are summarized in Tables 2 and 3. As before, all values are averages from 15 runs. First consider the final G values shown in Table 2. When the population is initialized to 80% #'s there is little difference among the three match scores. The only statistically significant differences are with pattern C3, where both

M2 and M3 do better than M1. This is interesting because C3 is a category that has no generalizations other than the set of all messages. M1 operates by seizing upon matching taxa quickly, then refining them to fit the situation. This strategy is frustrated when general taxa that consistently match are hard to find. Since M2 and M3 can both take advantage of other information, they do not have this problem with C3. When the population is initialized to 33% #'s the liabilities of M1 become very obvious. For each pattern category, the performance of M2 and M3 are both statistically significant improvements over M1.

In order to further understand the behavior of the match scores, we also compare them using DeJong's [1975] *on-line performance* criterion. On-line performance takes into account every new structure generated by the genetic algorithm, emphasizing steady and consistent progress toward the optimum value. The structures of interest here are populations as models of the pattern characteristic. The appropriate

Table 2 Comparison of Final G Values			
80% #'s			
Category	Match Score		
	M1	M2	M3
C1	7.83	10.30	7.76
C2	4.95	2.25	4.32
C3	5.98	1.42	0.97
50% #'s			
C1	10.28	8.17	6.96
C2	16.72	7.03	4.39
C3	13.67	8.67	9.13
33% #'s			
C1	12.25	8.06	5.19
C2	25.13	13.99	10.37
C3	36.57	11.41	7.28

Table 3 Comparison of On-line Performance			
80% #'s			
Category	Match Score		
	M1	M2	M3
C1	25.75	24.33	22.93
C2	14.06	11.26	13.45
C3	7.75	4.29	2.82
50% #'s			
C1	34.41	26.3	21.98
C2	27.09	20.22	17.81
C3	21.26	14.78	13.54
33% #'s			
C1	26.35	21.46	17.23
C2	35.3	26.75	24.64
C3	40.16	19.34	15.66

on-line measure is therefore given by $f(T) = (\frac{1}{T}) \sum_{t=1}^{t=T} G(T)$, where T is the number of generations observed and $G(t)$ is the G value for the tth generation. The on-line performance of the match scores is given in Table 3. When there are 80% #'s, the only statistically significant difference is the one between M3 and M1 on category C3. In the case of 50% #'s, the statistically significant differences occur on C1, where both M2 and M3 outperform M1; and, on C2, where only M3 does better than M1. Finally, in the difficult case of 33% #'s, the differences between M3 and M1 are all statistically significant. M2 is significantly better than M1 only on category C3.

Taken together, these results suggest that M3 is the best of the three match scores. It consistently gives the best performance over a broad range of circumstances. Figure 1 shows that, even in the case of 33% #'s, M3 reliably leads the genetic algorithm to the perfect model for all three categories. Using M3 should therefore enhance the ability of classifier systems to categorize messages.

How should a classifier system use M3 to identify relevant classifiers? The criterion for relevance using a score like M3 is centered around the idea of a variable threshold. The threshold is simply the number of mismatched taxon positions to be tolerated. Initially the threshold is set to zero and relevance is determined as with M1. If there are no matching classifiers, or not enough to fill the system's channel capacity, the threshold can be slowly relaxed until enough classifiers have been found. Note that this procedure is like the conventional one in that it clearly partitions the classifiers according to whether or not they are relevant to a message. This means that negated conditions in classifiers can be treated as usual; namely, a negated condition is satisfied only when it is not relevant to any message.

DISCOVERING MULTIPLE CATEGORIES

In developing the match score M3, we have enhanced the ability of the genetic algorithm to discover the defining characteristic for a given pattern category. What if there is more than one category to learn, or a single category with more than one defining characteristic? In this section we show how to modify the genetic algorithm to handle this more general case. First, two modifications are proposed for the way individuals are selected to reproduce and to be deleted. Then, the modified algorithm is shown to perform as desired.

An Ecological Analogy

The basic genetic algorithm is a reliable way to discover the defining characteristic of a category. When there is more than one characteristic in the environment, however, straightforward optimization of match scores will not lead to the best set of taxa. Suppose, for example, there are two categories given by the characteristics 11**...** and 00**...**. The ideal population for distinguishing these categories would contain the classifier taxa 11##...## and 00##...##; that is, two specialized sub-populations , one for each category. The genetic algorithm as described so far will treat the two patterns as one category and produce a population of taxa having good performance in that larger category. In this case, that means the taxon ####...## will be selected as the best way to categorize the messages. The problem is obvious. Requiring each taxon to match each message results in an averaging of performance that is not always desirable.

Various strategies have been proposed for avoiding this problem. When the number of categories is known in advance, the classifier system can be designed to have several populations of classifiers [Holland and Reitman,1978]; or, a single population with pre-determined partitions and operator

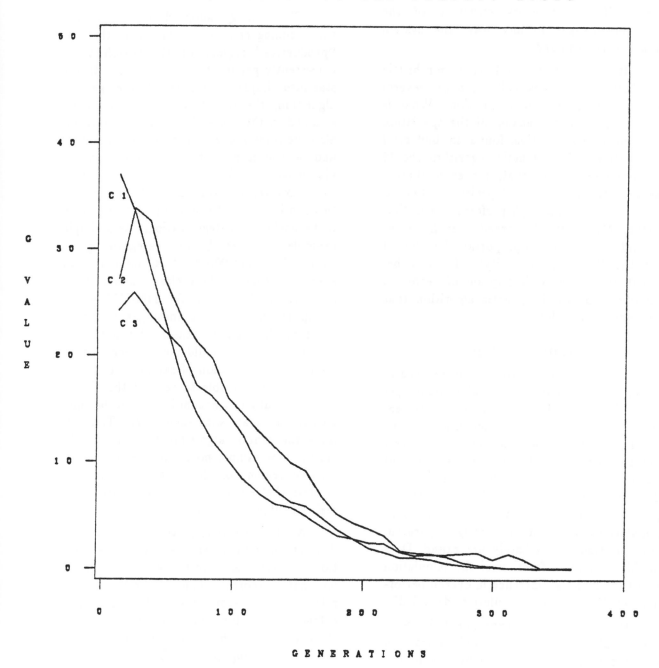

FIGURE 1

M3 CONVERGES TO THE PERFECT MODEL

restrictions [Goldberg,1983]. Both of these approaches involve building domain dependencies into the system that lead to brittleness. If the category structure of the domain changes in any way, the system must be re-designed.

It is preferable to have a non-brittle method that automatically manages several characteristics in one population. What is needed is a simple analog of the speciation and niche competition found in biological populations. The genetic algorithm should be implemented so that, for each characteristic or "niche" , a "species" of taxa is generated that has high performance in that niche. Moreover, the spread of each species should be limited to a proportion determined by the "carrying capacity" of its niche. What follows is a description of technical modifications to the genetic algorithm that implement this idea.

A Restricted Mating Strategy

If the genetic algorithm is to be used to generate a population containing many specialized sub-populations, it is no longer reasonable for the entire population to be modified at the same time. Only those individuals directly relevant to the current category need to be involved in the reproductive process. Given that the overall population size is fixed and the various sub-populations are not physically separated, two questions immediately are raised: Does modifying only a fraction of the population at a time make a difference in overall performance? How is a sub-population identified?

DeJong [1975] experimented with genetic algorithms in which only a fraction of the population is replaced by new individuals each generation. His results indicate that such a change has adverse effects on overall plan performance. The problem is that the algorithm generates fewer samples of the search space at a time. This causes the sampling error due to finite stochastic effects to become more severe. An increase in cumulative sampling error, in turn, makes it more likely that the algorithm will converge on some sub-optimal solution.

The strategy adopted here to reduce the sampling error is to make sure that the "productive" regions of the search space consistently get most of the samples. In the standard implementations of the genetic algorithm, the search trajectory is unconstrained in the sense that any two individuals have some non-zero probability of mating and generating new offspring (sample points) via crossover. This means, in particular, that taxa representing distinct characteristics can be mated to produce taxa not likely to be useful for categorization. As a simple example, consider the two categories given by 1111**** and 0000****. Combining taxa specific to each of these classes under crossover will lead to taxa like 1100**** which categorize none of the messages in either category. There is no reason why such functional constraints should not be used to help improve the allocation of samples. It therefore seems reasonable to restrict the ability of functionally distinct individuals to become parents and mate with each other. This will force the genetic algorithm to progressively cluster new sample points in the more productive regions of the search space. The clusters that emerge will be the desired specialized subpopulations.

As for identifying these functionally distinct individuals, any restrictive designation of parent taxa must obviously be based on match scores. This is because taxa relevant to the same message have a similar categorization function. Taken together, these considerations provide the basis for a *restricted mating policy*. Only those taxa that are relevant to the same message will be allowed to mate with each other. This restriction is enforced by using the set of relevant classifiers as the parents for each invocation of the genetic algorithm.

Crowding

Under the restricted mating policy, each set of relevant taxa designates a species. Each category characteristic designates a niche. Following this analogy, individuals that perform well in a given niche will proliferate while those that do not do well in any niche will become extinct. This ecological perspective leads to an obvious mechanism for automatically controlling the size of each sub-population. Briefly, and very simply, any ecological niche has limited resources to support the individuals of a species. The number of individuals that can be supported in a niche is called the *carrying capacity* of the niche. If there are too many individuals there will not be enough resources to go around. The niche becomes "crowded," there is an overall decrease in fitness, and individuals die at a higher rate until the balance between niche resources and the demands on those resources is restored. Similarly, if there are too few individuals the excess of resources results in a proliferation of individuals to fill the niche to capacity.

The idea of introducing a crowding mechanism into the genetic algorithm is not new. DeJong [1975] experimented with such a mechanism in his function optimization studies. Instead of deleting individuals at random to make room for new samples, he advocates selecting a small subset of the population at random. The individual in that subset most similar to the new one is the one that gets replaced. Clearly, the more individuals there are of a given type, the more likely it is that one of them will turn up in the randomly chosen subset. After a certain point, new individuals begin to replace their own kind and the proliferation of a species is inhibited.

A similar algorithm can be implemented much more naturally here. Because a message selects via match scores those taxa that are similar, there is no need to choose a random subset. Crowding pressure can be exerted directly on the set of relevant taxa. This can be done using the *strength* parameter normally associated with every classifier [Holland,1983]. The strength of a classifier summarizes its value to the system in generating behavior. Strength is continuously adjusted using the *bucket brigade* algorithm [Holland,1983] that treats the system like a complex economy. Each classifier's strength reflects its ability to turn a "profit" from its interactions with other classifiers and the environment. One factor bearing on profitability is the prevailing "tax rate". Taxation is the easiest way to introduce crowding pressure. Assume that a classifier is taxed some fraction of its strength whenever it is deemed to be relevant to a message. Assume, further, that all relevant classifiers share in a fixed sized tax rebate. The size of the tax rebate represents the limited resource available to support a species in a niche. When there are too many classifiers in a niche their average strength decreases in a tax transaction because they lose more strength than they gain. Conversely, when there are too few classifiers in a niche their average strength will increase. The crowding pressure is exerted by deleting classifiers in inverse proportion to their strength. The more individuals there are in a niche, the less their average strength. Members of this species are therefore more likely to be deleted. In a species with fewer members, on the other hand, the average strength will be relatively higher which means members are more likely to survive and reproduce. In this way, the total available space in the population is automatically and dynamically managed for every species. The number of individuals in a niche increases or decreases in relative proportion to the average strength in alternative niches.

Testing the New Algorithm

Having described the restricted mating policy and crowding algorithm, we now examine how well they perform in an actual implementation. The genetic algorithm used in previous experiments was modified as indicated above. The number of taxa in the population was increased to 200, and each taxon was given an initial strength of 320. A taxation rate of 0.1 was arbitrarily selected, and the tax rebate was fixed at 50*32; In other words, whenever there are 50 relevant taxa, the net tax transaction based on initial strengths is zero. Each generation the tax transaction is repeated 10 times to help make sure the strengths used for crowding are near their equilibrium values.

Four categorization tasks involving multiple characteristics were chosen to test the performance of the algorithm:

1) 11111111********
   00000000********

2) 11111111********
   ********11111111

3) 1111111111******
   ********11111111

4) 11111111********
   00000000********
   ********11111111

The first task involves two categories that are defined on the same feature dimensions. The second task contains categories defined on different dimensions. In the third task the categories share some relevant features in common. Finally, the fourth task involves three categories to be discriminated.

Experiments were performed on each of these tasks, running the genetic algorithm enough generations to produce 6000 new individuals per characteristic. Each generation, one of the characteristics was selected and a message belonging to that category was used to compute match scores. In the first three tasks, at least 50 relevant taxa were chosen per generation. Only 30 were chosen on task 4 to avoid exceeding the limited capacity of the population. All populations were initialized with 80% #'s. The results are summarized in Table 4 and show that the algorithm behaves as expected. The performance values are comparable to those obtained with M1 working on a simpler problem with a dedicated population. More importantly, an inspection of the populations revealed that they were partitioned into specialized sub-populations as desired.

CONCLUSIONS

This research has shown how to improve the performance of genetic algorithms in classifier systems. A new match score was devised that makes use of all of the information available in a population of taxa. This improves the ability of the genetic algorithm to discover pattern characteristics under changing conditions in

Table 4 Performance With Multiple Categories		
Task	On-line	Avg. *G* value for all categories
1	12.12	8.3
2	10.91	8.41
3	12.77	7.89
4	15.75	11.64

the environment. Modifications to the algorithm have been presented that transform it from a function optimizer into a sophisticated heuristic for categorization. The first modification, a restricted mating policy, results in the isolation and development of clusters of taxa, or sub-populations, correlated with the inferred structural characteristics of the pattern environment. The second modification, a crowding algorithm, is responsible for the dynamic and automatic allocation of space in the population among the various clusters. Together, these modifications produce a learning algorithm powerful enough for challenging applications. As evidence of this claim, a full-scale classifier system has been built along these lines that solves difficult cognitive tasks [Booker,1982].

Acknowledgements

† The ideas in this paper were derived from work done on the author's Ph.D. dissertation. That work was supported by the Ford Foundation, the IBM Corporation, the Rackham School of Graduate Studies, and National Science Foundation Grant MCS78-26016.

REFERENCES

Bethke, A.D. (1981), "Genetic Algorithms as Function Optimizers", Ph.D. dissertation, University of Michigan.

Booker, L.B. (1982), "Intelligent Behavior as an Adaptation to the Task Environment", Ph.D. dissertation, University of Michigan.

DeJong, K.A. (1975), "Analysis of the Behavior of a Class of Genetic Adaptive Systems", Ph.D. dissertation, University of Michigan.

Goldberg, D.E. (1983), "Computer-Aided Gas Pipeline Operation Using Genetic Algorithms and Rule Learning", Ph.D. dissertation, University of Michigan.

Hayes-Roth, F. (1973), "A Structural Approach to Pattern Learning and the Acquisition of Classificatory Power", Proceedings of the First International Joint Conference on Pattern Recognition, p. 343-355.

Hinton, G., Sejnowski, T., and Ackley, D. (1984), "Boltzmann Machines: Constraint Satisfaction Networks that Learn", Technical Report CMU-CS-84-119, Carnegie-Mellon University,.

Holland, J.H. (1975), *Adaptation in Natural and Artificial Systems*, University of Michigan Press, Ann Arbor.

Holland, J.H. (1976), "Adaptation" , In *Progress in Theoretical Biology 4* (Rosen, R. and Snell, F. eds). Academic Press, New York.

Holland, J.H. (1983), "Escaping Brittleness", Proceedings of the International Machine Learning Workshop, June 1983, Monticello, Illinois, pp.92-95.

Holland, J.H. and Reitman, J.S. (1978), "Cognitive Systems Based on Adaptive Algorithms", In *Pattern-Directed Inference Systems*, (Waterman, D. and Hayes-Roth, F. eds), pp. 313-329. Academic Press, New York.

Kuchinski, M.J. (1985), "Battle Management Systems Control Rule Optimization Using Artificial Intelligence", Technical Note, Naval Surface Weapons Center, Dahlgren, VA.

Kullback, S. (1959), *Information Theory and Statistics*, John Wiley and Sons, New York.

Multiple Objective Optimization with Vector Evaluated Genetic Algorithms

J. David Schaffer
Department of Electrical Engineering
Vanderbilt University
Nashville, TN 37235

ABSTRACT

Genetic algorithms (GA's) have been shown to be capable of searching for optima in function spaces which cause difficulties for gradient techniques. This paper presents a method by which the power of GA's can be applied to the optimization of multiobjective functions.

1. Introduction

There is currently considerable interest in optimization techniques capable of handling multiple non-commensurable objectives. Many practical problems are of this type where, for example, such factors as cost, safety and performance must be taken into account.

A class of adaptive search procedures known as genetic algorithms (GA's) have already been shown to possess desireable properties [3,10] and to out perform gradient techniques on some problems, particularly those of high order, with multiple peaks or with noise disturbance [4,5,6]. This paper describes an extension of the traditional GA which allows the searching of parameter spaces where multiple objectives are to be optimized. The software system implementing this procedure was called VEGA for Vector Evaluated Genetic Algorithm.

The next section of this paper will describe the basic GA and the vector extension. Then some properties are described which might logically be expected of this method. Some preliminary experiments on some simple problems are then presented to illuminate these properties and finally, VEGA is compared to an established multiobjective search technique on a set of more formidable problems.

2. A Vector Genetic Algorithm

Unlike many other search techniques which maintain a single "current best" solution and try to improve it, a GA maintains a set of possible solutions called a population. This population is improved by a cyclic two-step process consisting of a selection step (survival of the fittest) and a recombination step (mating). Each cycle is usually called a generation. More detailed descriptions of these operations may be found in the literature [3,4,5,6,10].

The question addressed here is, how can this process be applied to problems where fitness is a vector and not a scalar? How might survival of the fittest be implemented when there is more than one way to be fit? We exclude scalarization processes such as weighted sums or root mean square by the assumption that the different dimensions of the vector are non-commensurable.

When comparing vector quantities, the usual concepts employed are those proposed by Pareto [11,13]. For two vectors of the same size, the equality, less-than and greater-than relations require that these relations hold element by element. Another relation, partially-less-than, is defined as follows: vector $X = \{x1, x2, \ldots , xn\}$ is said to be partially-less-than vector $Y = \{y1, y2, \ldots , yn\}$ iff $xi \leq yi$ for all i and for at least one value of i, $xi < yi$. Assuming that minima are sought, if X is partially-less-than Y, then Y is said to be inferior to or dominated by X. The objective of a search for minima in a vector-valued space is, then, a search for the set of non-inferior members, or the members not dominated by any others. At least one member of this Pareto-minimal set will dominate each vector outside the set, but among themselves, none is dominated.

With these concepts in mind, a simple vector survival of the fittest process was implemented. The selection step in each generation became a loop, each time through the loop the appropriate fraction of the next generation was selected on the basis of another element of the fitness vector. This process, illustrated in figure 1, protects the survival of the best individuals on each dimension of performance and, simultaneously, provides the appropriate probabilities for multiple selection of individuals who are better than average on more than one dimension.

3. Some Anticipated Properties of VEGA

3.1 Multiple Solutions

One potential advantage of VEGA over other optimization searches should now be clear. Since the object of the search is a set of solutions, a GA has a built-in advantage by working with a population of test solutions. By comparing each individual in a population to every other, those who are dominated by any other/s can be flagged as inferior. The set of non-inferior individuals in each generation is the current best guess at the

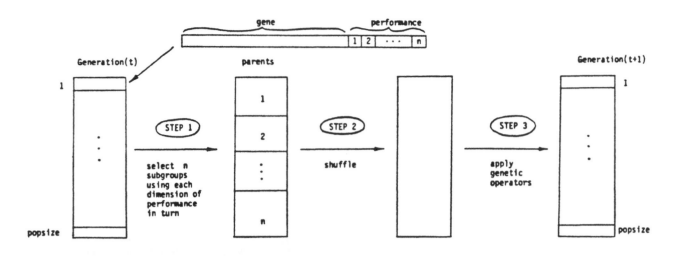

Figure 1. Schematic of VEGA Selection

Pareto-optimal (PO) set. By presenting a number of non-inferior solutions, VEGA provides the user with an idea of the tradeoffs required by his problem if a single solution must be selected. It should be noted that VEGA's view of non-inferiority is strictly local; it is limited to the current population. While a locally dominated individual is also globally dominated, the converse is not necessarily true. An individual who is non-dominated in one generation may become dominated by an individual who emerges in a later generation.

3.2 Possible Speciation

There is a potential problem with this vector selection process. Survival pressure is applied favoring extreme performance on at least one dimension of performance. If a utopian individual (i.e. one who excels on all dimensions of performance) exists, then he may be found by genetic combinations of extreme parents, but for many problems this utopian solution does not exist. For these problems, the location of the Pareto-optimal set or front is sought. This front will contain some members with extreme performance on each dimension and some with "middling" performance on all dimensions. Frequently, these compromise solutions are of most interest, but there may be danger of their not surviving VEGA's selection process. This might give rise to the evolution of "species" within the population which excel on different aspects of performance. This danger is expected to be more severe for problems with a concave PO front than for those with a convex one. See figure 2.

Two methods for combating this potential property of VEGA were conceived. One trick would be to provide a heuristic selection preference for non-dominated individuals in each generation. This would provide extra protection for the "middling" individuals.

Another, not necessarily exclusive, approach would be to try to encourage crossbreeding among the "species" by adding some mate selection

heuristics. In a traditional GA, mates are selected at random. On the assumption that utopian individuals are more likely to result from crossbreeding than inbreeding, such heuristics might speed the search.

4. Preliminary Experiments

4.1 The Test Functions

In order to test the properties of VEGA search, a set of three simple functions (f1, f2 & f3) was selected.

F1 was a single-valued quadratic function of three variables. (i.e. f1(x1,x2,x3) = x1**2 + x2**2 + x3**2). This function was run to test whether VEGA reverts to a traditional GA when the performance vector has only one dimension.

F2 was a two-valued function of one variable (i.e. f21(x) = x**2; f22(x) = (x-2)**2). The initial random population for the search on this function is illustrated in figure 3. In addition to the locations of x, f21 and f22, this figure also shows the dominated flag for each x (1 if dominated, 0 if not). The PO region is 0<=x<=2.

F3 was another two-valued function of one variable, but with two disjoint PO regions 0<=x<=2 and 4<=x<=5.

4.2 Heuristics

In order to mitigate the anticipated loss of "middling" individuals a heuristic was tested which gave an extra selection preference to locally non-dominated individuals. This preference took the form of numeric adjustments to the performance measures which were required by the selection algorithm to sum to zero across the population. Therefore, a small penalty was deducted from each inferior individual and the sum of these penalties was divided among the non-inferior individuals.

Experiments were also conducted to see if the search for the PO front could be improved by mate-selection heuristics which encouraged crossbreeding. Inbreeding, in this context, means a

94

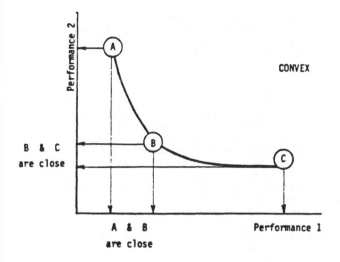

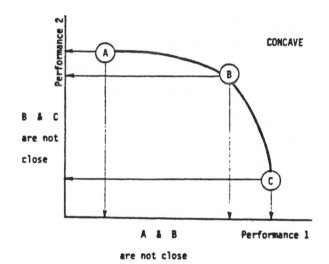

Figure 2. A Concave and Convex Pareto-Optimal Front

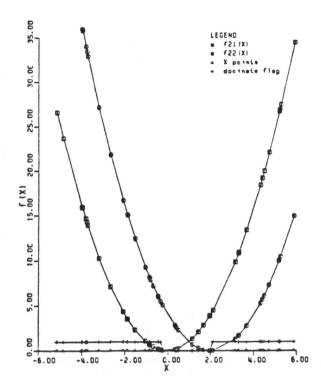

Figure 3. F2 Generation Zero

mating between two individuals whose high performance is on the same dimension. Two such heuristics were tested, both attempting to improve upon the performance of VEGA with random mating. Random mating was implemented by shuffling the population and mating pairs from the top, shown as step three in figure 1. Each heuristic proceeded by selecting a individual at random and then selecting a mate whose distance in performance space was maximum. Two distance measures were tested, Euclidian distance and "improvement" distance which was computed ignoring those dimensions on which the proposed mate performed worse.

4.3 Results

All of these experiments were conducted with populations of 30 individuals per dimension of the performance vector, and crossover and mutation rates of .95 and .01 respectively. This represents a smaller population size and higher rates of

application of the genetic operators than has been traditional [3,5]. These setting were, however, suggested by the work of Grefenstette [8].

On f1, VEGA replicated a search previously conducted on this function by a traditional GA [7] when started with the same random seeds. Thus, VEGA does appear to be a vector generalization of a scalar GA.

On f2, VEGA evolved the population illustrated in figure 4 in just three generations. While not all the individuals are in the PO region (0<=x<=2), those which are outside are known to be dominated. This result, combined with similar performance by VEGA on f3 yielded some confidence in the soundness of the VEGA approach.

However, during these experiments, a dangerous property of the heuristic selection preference for non-dominated individuals was discovered. It had a tendency to produce sudden premature convergence of the population to a suboptimal solution. This occurred when, in an early generation, only one or two individuals managed to be non-dominated. Then, the sum of the dominated penalties was large and, when divided among very few, gave them an overwhelming selection advantage. This lead to subsequent generations consisting only of offspring of a few parents with too little genetic diversity. After this observation, this heuristic was removed. VEGA has, so far, not exhibited the anticipated loss of the "middling" individuals from the PO set. Perhaps concave PO fronts are not a characteristic of many practical problems.

The mate-selection heuristics faired no better. Random mating proved superior to both of them. This was an encouraging finding for two-

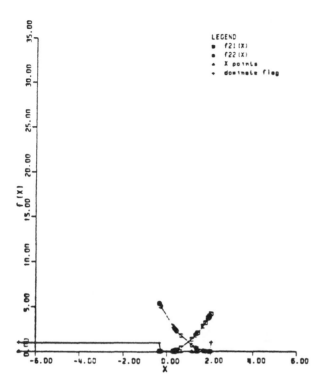

LEGEND
■ f21 (X)
■ f22 (X)
▲ X points
+ dominate flag

Figure 4. F2 Generation Three

valued problems, since the probability of inbreeding with random mating decreases as the number of dimensions of performance increases. All subsequent experiments utilized neither of these heuristics.

5. Comparison of VEGA with ARSO

Once some confidence was acquired that VEGA was able to conduct a genetic search in spaces with multiple objectives, it was desired to compare the performance of VEGA with that of an established technique for multiple objective search.

5.1 The ARSO Technique

For comparison purposes, the Adaptive Random Search Optimization (ARSO) procedure, pioneered by Beale [1,2], was selected. ARSO requests a starting point in the parameter space to be searched and proceeds to try to improve upon it by randomly perturbing the parameters. Statistics (mean & variance) are maintained for all perturbations which produce improvements (defined as a new solution which dominates the old one), and these statistics are used to guide the future perturbations. Random perturbation techniques have been shown to solve a large class of optimization problems faster than gradient techniques when the number of parameters exceeds four, and furthermore, the convergence time seems to increase only linearly with this number. ARSO had already exhibited high performance in problems of the sort tested here.

5.2 Some Methodological Problems

The comparison of two search procedures presents some methodological problems which are complicated when there are multiple objectives. One approach is to run each procedure until the solution is within some tolerance of a known solution and then compare the computational effort. This approach was rejected since the true solution was not known for the test problems. It was desired to compare the methods on problems whose solutions were not known so as to include in the comparison, the stopping criterion of each method. ARSO has a threshold on the number of perturbations tried without finding an improvement which forces a halt to the search. VEGA has no such preset stopping criterion and is stopped by the user when no further improvement is evident.

Another approach is to run both procedures for the same amount of computational effort and then compare the quality of the solutions. Comparing vector solutions is probably best done by checking if any are dominated by those provided by the other procedure. If not, then a tie must be declared. This approach may be unfair since ARSO reports only a single solution while VEGA may report several.

The approach adopted was to run each procedure to its natural stopping criterion. All proposed non-dominated solutions were then compared and, if any were found to be inferior, they were rejected. (Included in this set were solutions provided by Hartley [9] who used a variant of ARSO, but solved the scalar problem of the equally weighted sum of the errors on all dimensions.) Then, the number of "ultimately" non-dominated solutions found by each procedure was plotted against computational effort (number of function evaluations).

5.3 The Test Functions

A set of three problems, drawn from the domain of control engineering was contributed by a colleague, Hartley [9]. All involved the simulation of a system with a different integration operator for each of the system state variables. The systems were of orders 2, 3 and 7 respectively. The object of the search was an optimal set of integrators, each characterized by three parameters, making the dimensions of the parameter search spaces 6, 9 and 21, respectively. The performance measures were the rms error of the simulated solution from a known solution, one for each of the state variables making the dimensions of the performance spaces 2, 3 and 7.

All searches were conducted using the same GA parameters as were used for the preliminary problems. The integrator parameter sets were gray coded (see Schaffer[12] or Bethke[3]) to 12 bit precision, making the binary search spaces $2^{**}(12^*6) = 4.7 * 10^{**}21$, $2^{**}(12^*9) = 3.5 * 10^{**}32$ and $2^{**}(12^{**}21) = 7.2 * 10^{**}75$ for the three systems, respectively.

5.4 Results

While the true system behavior was assumed known, the object of the search was for optimal integrators for the simulations, and these were not known. Thus the problem of when to stop searching had to be faced. To illustrate, a scatter plot of performance of the initial random generation for the second order system is shown in figure 5. Figure 6 shows that considerable improvement had been achieved in three generations. Figure 7 shows the leading edge of the population after 49 generations. Note that the axes have been expanded three orders of magnitude. After running VEGA to generation 110 no substantial increase in performance was evident. See figure 8. There are however, several more points on what appears to be the PO front. Thus, a decision to stop such a search must be a judgement call based on a belief that the PO front has been located and that further search effort would be wasted.

The experiences were similar for the third and seventh order systems, but scatter plots for these high order systems could not be drawn.

Before proceding to the comparison of VEGA with ARSO, it may be instructive to illustrate one of the ARSO searches in the second order system problem. Figure 9 traces the improvements in the solution found by ARSO and is presented on the same axes scales used for figures 5 to 8. ARSO found a solution which was judged "ultimately" non-dominated in 607 evaluations. ARSO's stopping criterion halts if no improvement is located after 1000 consecutive evaluations and so this run continued until 1607 evaluations and halted.

A second run of ARSO was initiated with one of the two non-dominated individuals from the initial population generated by VEGA. This run halted after about 1300 evaluations, but its solution was inferior. VEGA, on the other hand, did not locate its first "ultimately" non-dominated solution until 2621 evaluations and by 6000 it had found eight. These results are shown in figure 10.

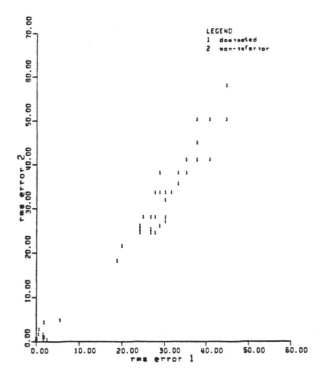

Figure 6. Second Order System — Generation Three

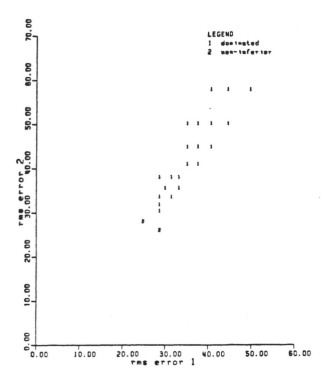

Figure 5. Second Order System — Generation Zero

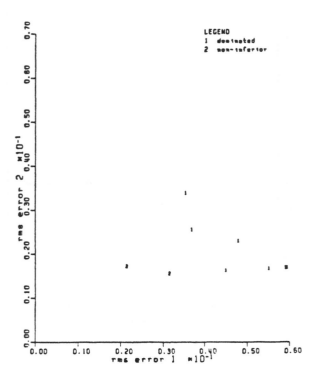

Figure 7. Second Order System — Generation 49

97

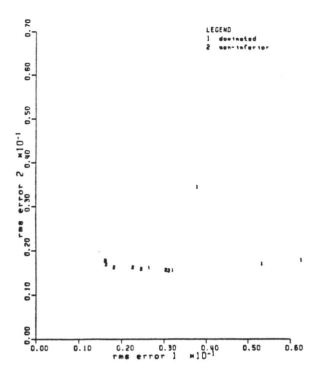

Figure 8. Second Order System — Generation 110

The tentative conclusion from these runs is that ARSO is fast, but may get trapped on local extrema. VEGA is slower, but more robust.

Four searches were made on the 3rd order system, two with VEGA and two with ARSO. VEGA was initiated with a random population for one run and given a whole population of clones of Hartley's solution for the other. ARSO was started with a reasonable starting point analogous to the starting point that lead to success on the 2nd order problem, and also the Hartley solution. The results of these tests are presented in figure 11. ARSO again found a good solution in under 2000 evaluations on its first run. When given a PO solution, it could only try for 1000 evaluations to improve it and then halt. VEGA found a non-dominated solution quite early in its search (415 evaluations), but because it was not sufficiently extreme on any one dimension of performance, it did not survive into future generations. More good solutions emerged later with VEGA having five after 10000 evaluations. VEGA, unlike ARSO, when given a PO solution, quickly located many variants of it.

The same four searches were run on the 7th order system. This time neither VEGA nor ARSO located any solutions which were not dominated by the Hartley solution. Hartley had used his knowledge of the problem to start his search at a close-to-PO point, but both VEGA and ARSO were started without this prior knowledge. The stopping criteria for ARSO had been relaxed to lengthen the search and a variance parameter had also been relaxed so as to broaden the search, but after almost 12000 evaluations no non-dominated solutions had been found. Its best solution at that time was

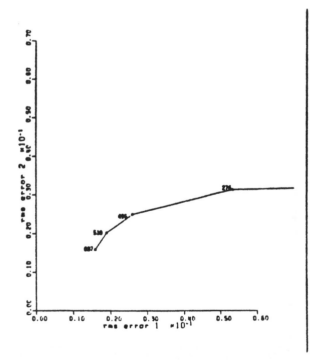

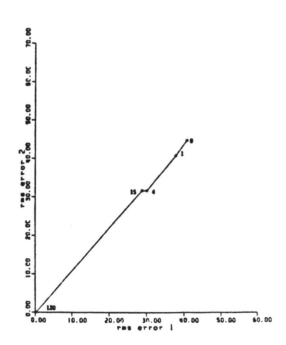

Figure 9. Trace of ARSO Search, Second Order System

also known to be unstable. VEGA searched for almost 36000 evaluations without locating any solutions not dominated by Hartley's, however many of them were stable. Again, when told where to look, VEGA generated many more PO solutions.

The tentative conclusion, then, seems to have been supported by the higher order searches.

6. Discussion

The major finding of this research was that vectorization of performance feedback and the selection process of a GA can be successfully done. This opens the domain of multiobjective optimization problems to the already established power of genetic search.

Heuristic modifications of the traditional method to give selection preference to non-dominated members of a population and to try to improve on random mating proved to be inferior to the traditional method. The possibility that VEGA may have a weakness in the central region of a concave PO front cannot be eliminated, but empirical evidence to date suggests that it may not be serious.

The comparisons of VEGA with ARSO contain no small amount of "apples versus oranges." The methods differ in the number of solutions presented and in the way their searches are normally halted. However, both contain stochastic elements, both conduct multidimensional search and both are halted when no further improvement is apparent. Both may be started with random information, or may take advantage of prior knowledge the user possesses about his search space. In the comparison runs VEGA

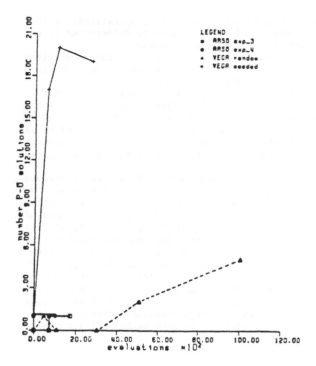

Figure 11. ARSO vs VEGA on Third Order System

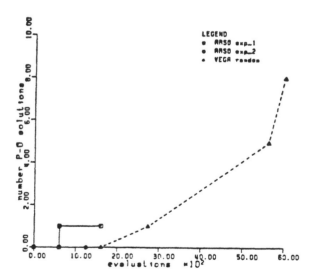

Figure 10. ARSO vs VEGA on Second Order System

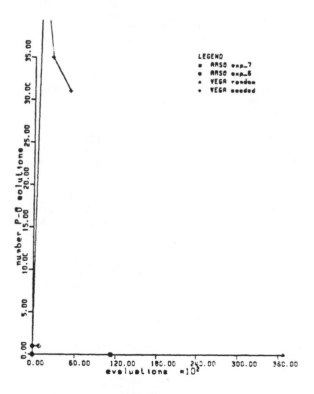

Figure 12. ARSO vs VEGA on Seventh Order System

99

was given several times more computational effort than was ARSO, due largely to differences in the methods for stopping each search.

The general conclusion of the comparison was that ARSO is capable of very quickly locating solutions to complex multidimensional problems, but its preformance may be less robust than VEGA's. VEGA, on the other hand, takes longer to locate the good regions of complex search spaces, but seems to be able to do so more reliably. This conclusion is not dissimilar to previous results from comparison of scalar genetic search with gradient techniques [5,6].

Finally, a simple method has been conceived which may improve both VEGA and ARSO. By maintaining a data structure "off to the side" containing all non-dominated solutions encountered in the search, VEGA would be protected against the loss of good but not extreme individuals, as occurred in the search on the 3rd order problem. Similarly, ARSO would then have the power to report a number of solutions instead of only one. Furthermore, by monitoring the adding and subtracting of members to this set, both techniques might be given a more rational stopping criterion. Work on this addition to both methods will commence in the near future.

REFERENCES

1. Guy O. Beale Optimal Aircraft Simulator Development by Adaptive Random Search Optimization, Ph.D. Dissertation, University of Virginia, Charlottesville, Virginia, May 1977.

2. Guy O. Beale and Gerald Cook, "Optimal digital simulation of aircraft via random search techniques," J. Guidance and Control, Vol. 1, no. 4, July-Aug. 1978.

3. Albert Donally Bethke Genetic Algorithms as Function Optimizers, Ph.D. Dissertation, University of Michigan, Ann Arbor, Michigan, Jan 1981.

4. Anne Brindle Genetic Algorithms for Function Optimization, Ph.D. Dissertation, University of Alberta, Edmonton, Alberta, Canada, Jan 1981.

5. Kenneth DeJong, Analysis of the behavior of a class of genetic adaptive systems, Ph. D. Dissertation, University of Michigan, Ann Arbor, Mich., 1975.

6. Kenneth DeJong, "Adaptive System Design: A Genetic Approach," IEEE Trans. Systems, Man and Cybernetics, Vol. SMC-10 No. 9, Sept 1980.

7. John Grefenstette, "A user's guide to GENESIS," Tech. Report CS-83-11, Computer Science Dept., Vanderbilt University, Nashville, Tenn., Aug. 1983.

8. John Grefenstette, "Genetic algorithms for multilevel adaptive systems," IEEE Trans. on Systems, Man and Cybernetics, in press.

9. Thomas Hartley, Parellel Methods for the Real Time Simulation of Still Nonlinear Systems, Ph.D. Dissertation, Vanderbilt University, Nashville, Tenn., Aug. 1984.

10. John H. Holland, Adaptation in Natural and Artificial Systems, University of Michigan Press, Ann Arbor, Michigan 1975.

11. V. Pareto, Cours d'Economie Politique, Rouge, Lausanne, Switzerland, 1896.

12. J. David Schaffer Some Experiments in Machine Learning Using Vector Evaluated Genetic Algorithms, Ph.D. Dissertation, Vanderbilt University, Nashville, Tennessee., Dec 1984.

13. Thomas L. Vincent and Walter J. Grantham, Optimality in Parametric Systems, John Wiley and Sons, New York, 1981.

Adaptive Selection Methods for Genetic Algorithms

James Edward Baker

Computer Science Department
Vanderbilt University

Abstract

Premature convergence is a common problem in Genetic Algorithms. This paper deals with inhibiting premature convergence by the use of adaptive selection methods. Two new measures for the prediction of convergence are presented and their accuracy tested. Various selection methods are described, experimentally tested and compared.

1. Introduction

In Genetic Algorithms, it is obviously desirable to achieve an optimal solution for the particular function being evaluated. However, it is not necessary or desirable for the entire population to converge to a single genotype. Rather the population needs to be diverse, so that a continuation of the search is possible. The loss of an allele indicates a restriction on the explorable search space. Since the full nature of the function being evaluated is not known, such a restriction may prevent the optimal solution from ever being discovered. If convergence occurs too rapidly, then valuable information developed in part of the population is often lost. This paper deals with the control of rapid convergence.

Three measures are typically used to compare genetic algorithms. They are: the *Online Performance*, the average of all individuals that have been generated; the *Offline Performance*, the average of the Best Individuals from each generation; and the *Best Individual*, the best individual that has been generated. We attempt to optimize functions and therefore use the Best Individual measure for comparison. In order to improve this measure, we promote diversity within the population and control rapid convergence. Increased diversity detrimentally affects the Online Performance measure and inhibited convergence detrimentally affects the Offline Performance measure. Improving these two performance measures is not in the scope of this paper.

Methods for the prediction of a rapid convergence are the topics of section 2. Section 3 will describe various algorithms with which to slow down convergence, and section 4 will present their results. A conclusion section will follow the results.

2. Prediction of Rapid Convergence

There are two different aspects to the control of rapid convergence. First, how can one tell that it has occurred and second, how can one predict when it will occur.

Recognizing rapid convergence after it has occurred is rather straightforward. By its very meaning, a rapid convergence will result in a dramatic rise in the number of lost and converged alleles. A *lost allele* occurs whenever the entire population has the same value for a particular gene. Thus subsequent search with that gene is impossible. A *converged allele*, as defined by DeJong [1], is a gene for which at least 95% of the population has the same value. However, the effects of rapid convergence are not limited to only those alleles which are indicated by these measures. A rapid take over of the population will cause all genes to suddenly lose much of their variance. We define *bias* as the average percent convergence of each gene. Thus for binary genes, this value will range between 50, for a completely uniform distribution (in which for each gene there are as many individuals with a one as a zero) and 100, for a totally converged population (in which each gene has converged to a one or a zero). The bias measure provides an indication of the entire population's development without the disadvantage of a threshold, such as the one suggested by DeJong to indicate a converged allele. A threshold does not indicate the amount by which individuals exceed it or the number of individuals which fall just short. We can therefore monitor the sudden jumps in the lost, converged or bias values to determine when a rapid convergence has occurred.

The prediction of rapid convergence is necessary for selection algorithms to be able to adapt accordingly. Lost, converged or bias values cannot be used for this purpose since their measurement occurs after potentially vital information has been discarded. Two different prediction methods will be described.

A common cause of rapid convergence is the existence of *super individuals*, that is individuals which will be rewarded with a large number of offspring in the next generation. Since the population size is typically kept constant, the number of offspring allocated to a super individual will prevent some other individuals from contributing any offspring to the next generation. In one or two generations, a super individual and its descendants may eliminate many desirable alleles which were developed in other individuals. The first method addresses the problem of super individuals by setting a threshold on an individual's expected number of offspring. If, after all individuals in a generation have been evaluated, an individual has an expected value above this threshold, then a rapid convergence is deemed imminent.

A closer analysis of rapid convergence leads to the second measure. Rapid convergence is not caused solely by an individual receiving too many offspring, but also by the often related situation of many individuals being denied any offspring. Thus rapid convergence may also be predicted not by monitoring how many offspring the most fortunate individual receives, but rather by monitoring how many individuals receive no offspring. We define *percent involvement* as the percentage of the current population which contributes offspring for the next generation. The percent involvement measure has the advantage that it can predict a rapid convergence caused by several individuals even when none of them would exceed the threshold of the first method. If that threshold were lowered to catch this case, then it might predict a rapid convergence when it was not occurring.

3. Modifications to Selection

This section presents various methods designed to prevent or control rapid convergence. There are basically two choices: either develop a fixed selection algorithm which avoids rapid convergence; or develop a hybrid system, which adapts its selection algorithm to handle rapid convergence when it occurs.

3.1. Standard Selection

The expected value model presented by DeJong is taken as the standard for comparison, since a number of properties about its behavior have been proven [1,3]. This model evaluates each individual and normalizes the value with respect to the population's average. The result for each individual is called his expected value and determines the number of offspring that he will receive. In our implementation [2], the actual number of offspring will be either the floor or the ceiling of the expected value. Thus, the number of offspring attributed to a particular individual is approximately directly proportional to that individual's performance. This direct proportionality is necessary for Holland's theorems to hold, however it is also the core of this method's susceptibility to rapid convergence. Since there is no constraint on an individual's expected value, an individual can have as many offspring as the population size will allow. Therefore, the expected value model can exhibit rapid convergence leading to sharp increases in lost, converged and bias values as well as non-optimal final results.

Because of the theoretical advantages associated with the expected value model, all of the hybrid systems listed below use this selection algorithm when rapid convergence is not indicated.

3.2. Ranking

One way to control rapid convergence is to control the range of trials allocated to any single individual, so that no individual receives many offspring. The *ranking system* is one such alternative selection algorithm. In this algorithm, each individual receives an expected number of offspring which is based on the rank of his performance and not on the magnitude. There are many ways to assign a number of offspring based on ranking, subject to the following two constraints:

1. the allocation of trials should be monotonically increasing with respect

to increasing performance values, to provide for desirable rewarding;

2. the total of the individual allocation of trials should be such that the desired number of individuals will be in the next generation.

Determining the values for our ranking experiments was done by taking a user defined value, MAX, as the upper bound for the expected values. A linear curve through MAX was taken such that the area under the curve equaled the population size. For this construction, several values are easily derivable:

1. lower bound,
   MIN = 2.0 - MAX ;

2. difference between "adjacent" individuals,
   INC = 2.0 * (MAX - 1.0) / Population Size ;

3. lowest individual's expected value,
   LOW = INC / 2.0 .

Hence for a population size of 50 and a MAX of 2.0 : MIN = 0.0, INC = 0.04, LOW = 0.02.

However, ranking with MAX = 2.0 causes the population to be driven to converge during every generation, including very stable searching periods, i.e., all individuals being within 10% of the mean. Ranking forces a particular percent involvement rather than preventing low percent involvement values from occurring. Our experiments show the desirable range for the percent involvement value is between 94% and 100%. The above settings force a percent involvement value of approximately 75%, and hence cause undesirable convergence. To prevent this, one must choose a MAX value of approximately 1.1, which force the percentage involvement value into the desirable range.

3.3. Hybrid Systems

The following two systems use the expected value model as the default. When rapid convergence is predicted, the system will temporarily switch to a different selection method, designed to better handle the situation.

3.3.1. Hybrid with Ranking

We have investigated two systems in which Ranking is used as the alternative selection method. The systems differ only in the way in which a rapid convergence is predicted. The first one uses a threshold on the maximum allowable expected value; the second system uses a threshold on the minimum allowable percent involvement. Ranking was chosen since, as described previously, it should work better during periods of rapid convergence and the expected value model should work better during the other periods. Thus, these two system's strengths and weaknesses complement each other and should create a good hybrid.

3.3.2. Dynamic Population Size

Recall that one cause of rapid convergence is that super individuals prevent other individuals from having offspring. This is due to the enforcement of a constant population size, and clearly results in a drop in the percent involvement. If the population size were allowed to grow, then a super individual would not force the elimination of many other individuals.

The *dynamic population size* method is implemented by enforcing a lower bound on the percent involvement. This is done by adding individuals to the population until both the original population size and the acceptable value for the percentage involvement are reached. Due to the requirements of crossover, additions to the population are made in increments of two. Additional individuals are added to the system on the same basis as before, that is by their expected value.

During periods of slow convergence, the size of the population will be constrained toward the original population size, since the lower bound on the percent involvement will be satisfied before the entire current population is chosen. Although the population size may grow as large as deemed necessary (within physical memory limitations), it will be guided back to the original population size during periods of slow convergence, as long as the lower bound value is set below 100%. Furthermore, on our system, the floor of the individual's expected value was taken as a minimum for his number of offspring. This periodically led to percent involvement values which were higher than the required lower bound even during a time of high population size. This is a characteristic of our implementation and has not yet been investigated for its desirability.

This method has good intuitive appeal and has the advantage of using the expected value model throughout. The advantage of reacting differently to differing magnitudes of potential rapid convergence is also present.

A possible disadvantage of this system is that a super individual can still obtain a large percentage of the population very quickly; while other individuals are not completely lost, their effect on the population is tremendously undermined.

4. Results

All experiments were performed using the Genesis System [2] at Vanderbilt University. The initial population size was set to 50, the crossover rate to 0.6, and the mutation rate to 0.001. Each curve in figures 1 through 10 were taken from single, representative executions of the appropriate functions. Each curve in figures 11 and 12 represent the average of five executions.

4.1. Detection

In order to confirm the predicting capability of the percent involvement and greatest expected value, a function was designed on which a standard GA would experience a rapid convergence. This function had a gentle slope over more than 99.5% of the search space. A steep, highly valued spike existed in the remaining one half of one percent. To achieve the optimal result, the system needed to find the spike and then to optimize within it. The outlying, gentle slope discouraged those alleles necessary for the optimal result. Thus when a super individual occurred, that is one within the spike, vital information was likely to be lost.

This function was used with the expected value model of selection. For each generation, the values of the percent involvement and greatest expected value were output. For these values to be useful as predictors, they must noticeably change prior to a rapid increase in the lost,

converged and bias values. Graphs comparing these predictors with the lost, converged and bias values can be seen in figures 1 - 6.

Figure 1 shows that the percent involvement value drops sharply, prior to the dramatic rise in the lost value. A similar relationship exists for the converged value in Figure 2. Figure 3 shows the bias value climbing before the percent involvement value has reached its minimum. This occurs since the percent involvement values are already below their normal range prior to reaching its minimum. The normal range for the percent involvement has been found to be between about 94% and 100% . The drop in the percent involvement which occurs around the 23rd generation causes no appreciable effect, since (as the bias value indicates) the population is already over 90% converged. Note the first indication that a rapid convergence has occurred is given by the bias value, and the last indication by the lost value. This is seen by the primary increase occurring in the fourth generation for the bias value, in the fifth generation for the converged value and in the sixth generation for the lost value.

The maximum expected value also experienced a sharp change prior to the rapid convergence. Figures 4, 5 and 6 show this clearly. Note that the spike in the maximum expected value occurred one generation before the minimum percent involvement value. This shows clearly the more global nature of the percent involvement discussed earlier. That is for generation number three, there was a single super individual, evident from the maximum expected value and percent involvement. However, the largest loss of other members of the population occurred during the following generation, when this super individual's offspring were reproducing. This is seen in the fourth generation's percent involvement value.

Figures 1 - 6 show that both the percent involvement and the maximum expected value provide a good prediction of the occurrence of rapid convergence in this example. However, the percent involvement appears to be superior in general, since it can detect some rapid convergence not caused by a super individual.

4.2. Comparison of Methods

Recall that a superior method potentially produces better Best Individuals by retaining diversity in the population and controlling rapid convergence. Thus the lost, converged and bias values should remain low with increases occurring only gradually.

The various selection methods discussed in section 3 were tested with a variety of functions. In all cases, Ranking uses a MAX of 1.1, as discussed in section 3.2. Figure 7 compares their lost alleles values for the same function used for Figures 1 - 6. These results were chosen because they are fairly representative of the various functions tested. The standard selection method consistently experienced rapid convergence sooner and more dramatically. This can also be seen in Figure 8, a comparison of the bias values for the same function. Just as consistently, the ranking system did not rapidly converge.

Figures 9 and 10 show the loss of alleles and bias for Shekel's "foxhole problem" studied by DeJong [1,4]. These figures also show the same two characteristics: 1) standard selection performing worst; and 2) ranking performing best. However, note the vast superiority of the hybrid system which was based on the percent involvement over the other hybrid system. This is probably the result of the percent involvement's superior ability to predict rapid convergence.

The population variance method performed no worse than the standard system, but for this function, it did not perform significantly better. The hybrid methods did experience convergence. However, it is delayed, it is extended over more generations, and its magnitude is lessened. Therefore one should expect these systems to be able to produce better final solution than the standard system, although it may take longer.

Figures 11 and 12 show the average Best Individual versus trial number for the various methods. These values represent the average for five executions. Figure 11 is for the "foxhole problem". Note that all of the methods performed at least as well as standard selection, given a sufficient number of trials. Furthermore, standard selection has lost nearly half of its

alleles by the 2000th trial. Ranking has outperformed all the others and after 2000 trials has not lost any alleles. Thus ranking has the best final result and the best potential for improvement. Note that ranking is the slowest of the methods, not producing a "competitive" result until about the 900th trial. This causes the Offline Performance to be very bad, and the high diversity causes the Online Performance to suffer.

Figure 12 is from a function which has a sharp optimal region which the system must find. The function also has various local optima which may cause convergence before the discovery of this region. Figure 12 indicates the relative ability of the systems to find the optimal region. Each curve plotted represents the average over five executions. Ranking was able to find the region four out of five times, but again was the slowest in starting.

5. Conclusions

To varying degrees, all of the methods discussed in this paper were able to control rapid convergence. The ranking method shows the greatest promise. It results in better solutions for many functions experiencing rapid convergence and it maintains virtually all of its alleles. This gives it the potential for continued search and even further improvement on its solutions. The primary drawback to this system is that it requires a larger number of trials to obtain these results, especially for functions not exhibiting rapid convergence. We have observed that in the expected value model all of the individuals are typically within 20% of the mean during non-rapidly converging periods. Hence, the ranking system, with MAX = 1.1 , should be roughly equivalent to the expected value system during these periods, yet has the advantage of being able to control rapid convergence. Of course, it also has the disadvantage of avoiding rapid convergence, even when the convergence is desirable. Hence, ranking warrants further study both for its robustness and its particular handling of rapid convergence.

At present there is insufficient justification to rank the other methods' performance. However, they all represent improvements over the standard selection algorithm. For some functions they were able to significantly slow down the rapid convergence and retain more diversity within their genes. This typically led to better final results than the standard selection algorithm would produce. Standard selection did not outperform any of the other methods on any of the functions tested, given a sufficient number of trials.

Both the percent involvement and the expected value provide a prediction of rapid convergence and can be used to help control it. However, the prediction based on the expected value applies only for rapid convergence caused by super individuals. Therefore, the percent involvement value should be used as a more general predictor.

Many other methods to control rapid convergence should be studied and compared. Among them are:

1. providing a simple upper bound on the number of offspring allowable to an individual;

2. limit super individuals to a very small number of offspring (1 or 2) in combination with the elitist strategy;

3. a ranking system based on non-linear curves or adaptively changing curves;

4. providing multiple thresholds for the varying population size method, to use separate thresholds for growing and for shrinking the population, or separate values for the prediction and the processing phases.

The methods presented in this paper should be tested further on a larger number of functions before definitive conclusions can be made.

References

1. K. A. DeJong, *Analysis of the behavior of a class of genetic adaptive systems*, Ph.D. Thesis, Dept. Computer and Communication Sciences, Univ. of Michigan (1975).

2. J. J. Grefenstette, "A user's guide to GENESIS," Tech. Report CS-83-11, Computer Science Dept., Vanderbilt Univ. (August 1983).

3. J. H. Holland, *Adaptation in Natural and Artificial Systems,* Univ. Michigan Press, Ann Arbor (1975).

4. J. Shekel, "Test functions for multimodal search techniques," *Fifth Ann. Princeton Conf. Inform. Sci. Syst.* (1971).

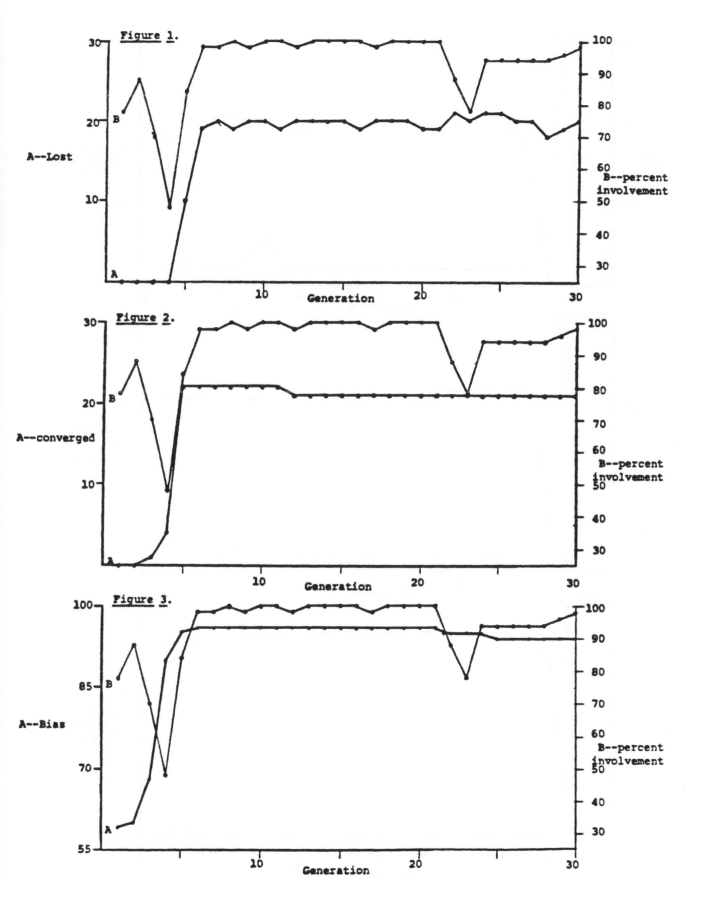

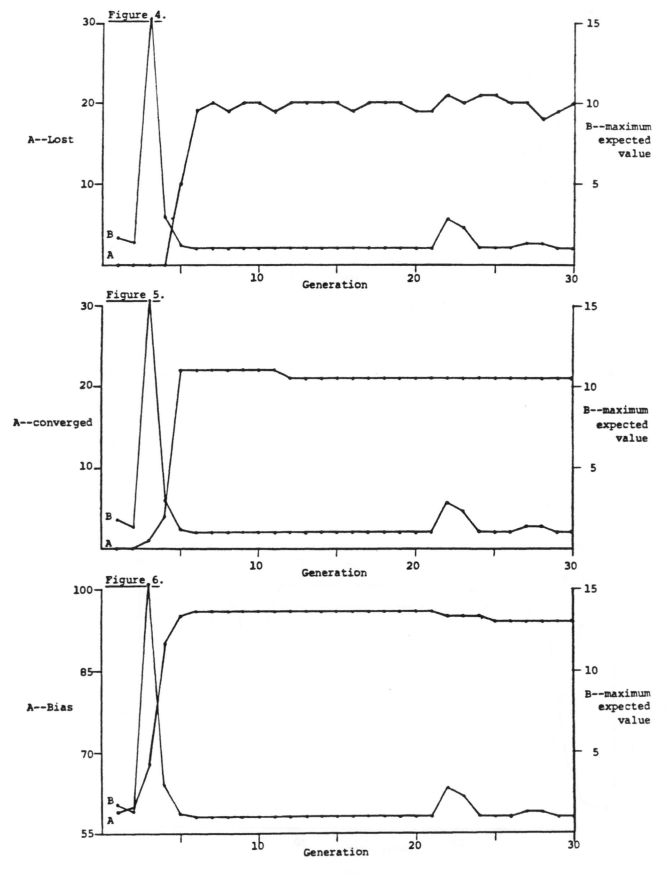

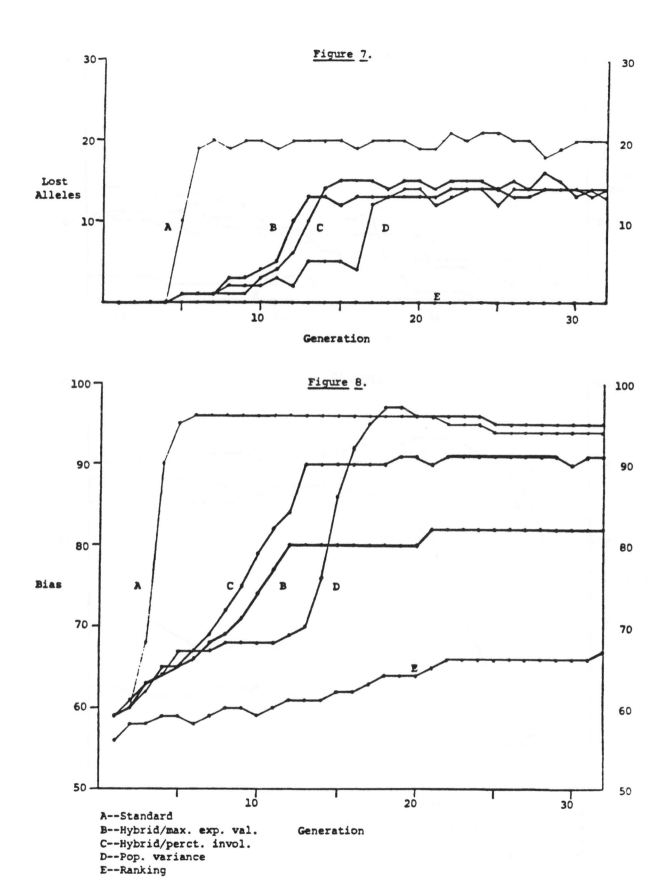

Figure 7.

Lost Alleles

A B C D

E

Generation

Figure 8.

Bias

A C B D

E

Generation

A--Standard
B--Hybrid/max. exp. val.
C--Hybrid/perct. invol.
D--Pop. variance
E--Ranking

109

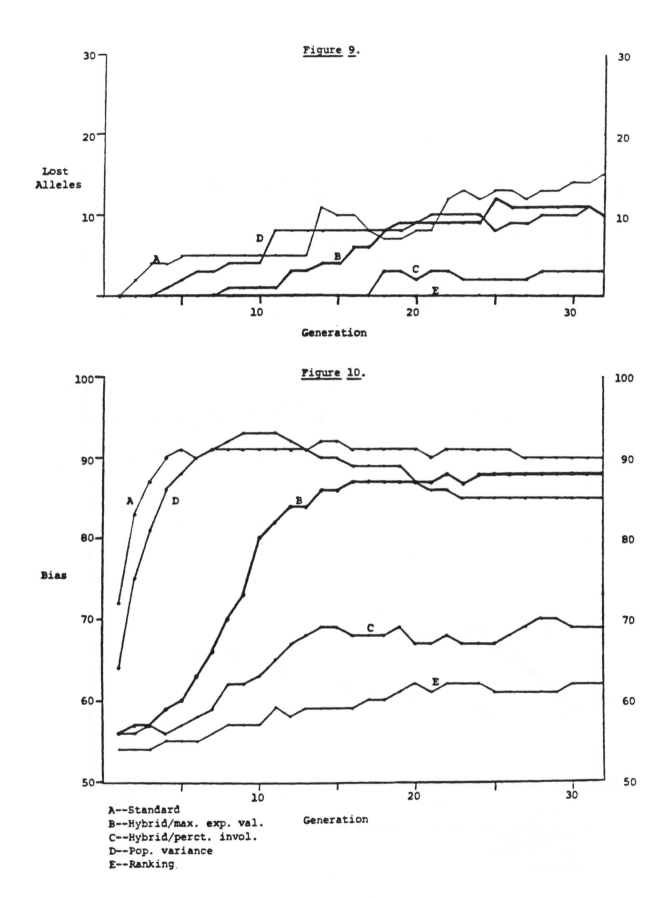

Figure 9.

Figure 10.

A--Standard
B--Hybrid/max. exp. val.
C--Hybrid/perct. invol.
D--Pop. variance
E--Ranking.

110

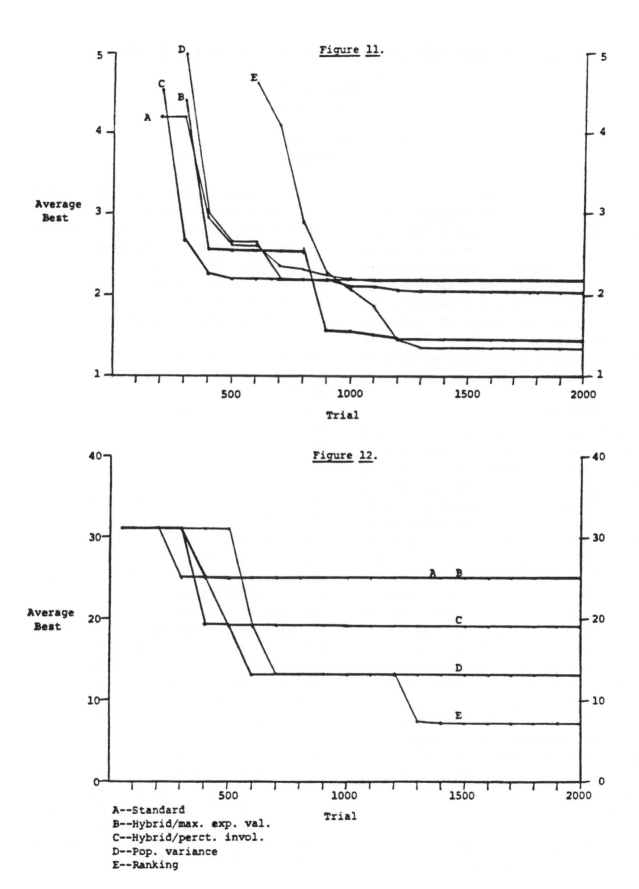

Figure 11.

Figure 12.

A—Standard
B—Hybrid/max. exp. val.
C—Hybrid/perct. invol.
D—Pop. variance
E—Ranking

Genetic Search with Approximate Function Evaluations

John J. Grefenstette[1]
J. Michael Fitzpatrick

Computer Science Department
Vanderbilt University

Abstract

Genetic search requires the evaluation of many candidate solutions to the given problem. The evaluation of candidate solutions to complex problems often depends on statistical sampling techniques. This work explores the relationship between the amount of effort spent on individual evaluations and the number of evaluations performed by genetic algorithms. It is shown that in some cases more efficient search results from less accurate individual evaluations.

1. Introduction

Genetic algorithms (GA's) are direct search algorithms which require the evaluation of many points in the search space. In some cases the computational effort required for each evaluation is large. In a subset of these cases it is possible to make an approximate evaluation quickly. In this paper we investigate how well GA's perform with approximate evaluations. This topic is motivated in part by the work of De Jong [5], who included a noisy function as part of his test environment for GA's, but did not specifically study the implications for using approximate evaluations on the efficiency of GA's. Our main question is: Given a fixed amount of computation time, is it better to devote substantial effort to getting highly accurate evaluations or to obtain quick, rough evaluations and run the GA for many more generations? We assume that the evaluation of each structure by the GA involves a Monte Carlo sampling, and the effort required for each evaluation is equal to the number of samples performed.

Since the GA's we consider do not obtain accurate evaluations during the search, the traditional metrics, *online performance* and *offline performance*, are not appropriate (or at least not easily obtained). Instead, we assume that the GA runs for a fixed amount of time, after which it yields a single answer. The performance measurement we use is the *absolute performance*, that is, the exact evaluation of the suggested answer after a fixed amount of time.

In section 2 we describe the statistical evaluation technique. In section 3 we describe the result of testing on a simple example evaluation function. In section 4 we describe the result of testing on image comparison functions. In section 5 we present future directions of research on approximate evaluations.

2. The Statistical Evaluation Technique

In this work we investigate the optimization of a function $f(x)$ whose value can be estimated by sampling. The variable x ranges over the space of structures representable to the GA. We are interested in functions for which an exact evaluation requires a large investment in time but for which an approximate evaluation can be carried out quickly. Examples of such functions appear in the evaluation of integrals of complicated integrands over large spaces. Such integrals appear in many applications of physics and engineering and are commonly evaluated by Monte Carlo techniques [13,15]. An example from the field of image processing, examined in detail below, is the comparison of two digital images. Here the integrand is the absolute difference between image intensities in two images at a given point in the image and the space is the area of the image.

---

[1]Research supported in part by the National Science Foundation under Grant MCS-8305693.

Throughout our discussions it is convenient to treat the function, $f(x)$, to be optimized as the mean of some random variable $r(x)$. In terms of the evaluation of an integral by the Monte Carlo technique, $f(x)$ would be the mean of of the integrand's value over the space and $r(x)$ is simply the set of values of the integrand over the space. The approximation of $f(x)$ by the Monte Carlo technique proceeds by selecting n random sample from $r(x)$. The mean of the sample serves as the approximation and to the extent that the samples are random, the sample mean is guaranteed by the law of large numbers to converge to $f(x)$ with increasing n. Once $f(x)$ is approximated, the desired value of the integral can be approximated by multiplying the approximation of $f(x)$ by the volume of the space. There are many approaches to improving the convergence of the sample mean and the confidence in the means for a fixed n [15]. We will not investigate these approaches. Here we will be concerned only with the sample mean and an estimate of our confidence in that mean.

The idea which we are exploring is to use as an evaluation function in the GA optimization of $f(x)$, not $f(x)$ itself, but an estimate, $e(x)$, of $f(x)$ obtained by taking n randomly chosen samples from $r(x)$. It is intuitive that $e(x)$ approaches $f(x)$ for large n. From statistical sampling theory it is known that if $r(x)$ has standard deviation $\sigma(x)$ then the standard deviation of the sample mean, $\sigma_s(x)$, is given by

$$(1) \quad \sigma_s(x) = \sigma(x)/\sqrt{n}$$

In general $\sigma(x)$ will be unknown. It is simple, however, to estimate $\sigma(x)$ from the samples using the unbiased estimate,

$$(2) \quad \sigma_e^2(x) = \sum_{i=1}^{n} (s_i - e(x))^2 / (n-1)$$

It is clear from equation (1) that reducing the size of $\sigma_s(x)$ can be expensive. Reducing $\sigma_s(x)$ by a factor of two, for example, requires four times as many samples. It is intuitive that the GA will require more evaluations to reach a fixed level of optimization for $f(x)$ when $\sigma_s(x)$ is larger. Concomitantly, it is intuitive that the GA will achieve a less satisfactory level of optimization for $f(x)$ for a fixed number of evaluations when $\sigma_s(x)$ is larger. What is is not obvious is which effect is more important here, the increase in the number of evaluations required or the increase in the time required per evaluation. The following experiments explore the relative importance of these two effects.

3. A Simple Experiment

As a simple example function we have chosen to minimize

$$f(x,y,z) = x^2 + y^2 + z^2$$

We imagine that $f(x,y,z)$ is the mean of some distribution which is parameterized by x, y and z, but instead of actually sampling such a function to achieve the estimate $e(x,y,z)$, we use

$$e(x,y,z) = f(x,y,z) + noise$$

where $noise$ represents a pseudo-random function chosen to be normally distributed and to have zero mean. The standard deviation, $\sigma_s(x,y,z)$, of $e(x,y,z)$ is in this case equal to that of the noise function and it is chosen artificially. No actual sampling is done. The advantage of this experimental scheme is that we can investigate the effects of many different distributions and sample sizes for each $\sigma_s(x,y,z)$ we choose without performing all the experiments.

In order to get some idea of the effect of the dependence of $\sigma_s(x,y,z)$ on x, y and z, we perform two different sets of experiments on $f(x,y,z)$: (a) $\sigma_s(x,y,z)$ independent of x, y and z; and (b) $\sigma_s(x,y,z) = \lambda_s f(x,y,z)$. The search space is limited to x, y and z between -5.12 and +5.12 digitized to increments of 0.01. The GA parameters are the standard ones suggested by De Jong [5]: population size 50, crossover rate 0.6, mutation 0.001. For the experiments of type (a) we determine for several values of σ_s the number of evaluations necessary to find x, y and z such that $f(x,y,z)$ falls below a threshold of 0.05. For the experiments of type (b) we determine for several values of λ_s the number of evaluations necessary to achieve the 0.05 threshold.

The results of 50 runs at each setting are shown in Figure 2. It is immediately obvious that these graphs are approximately linear. In fact linear regression analysis produces a correlation coefficient of 0.99 in each case. The linearity of these graphs simplifies their analysis considerably. To see the relative importance of number of evaluations versus time per evaluation we can start with the equation for the straight lines:

$$(3) \quad E_a = 1244 + 2268\sigma_s$$
$$(4) \quad E_b = 18{,}020 + 9285\lambda_s$$

where E_a and E_b are the number of evaluations required for case a and b, respectively. We imagine that the evaluations were obtained by sampling from a normal distribution whose standard deviation is σ in case (a) and $\lambda f(x,y,z)$ in case (b). In that case we can use Equation (1) for $\sigma_s(x,y,z)$ in both Equations (3) and (4) to get

$$(5) \quad E_a = 1244 + 2268\sigma/\sqrt{n}$$
$$(6) \quad E_b = 18{,}020 + 9285\lambda/\sqrt{n}$$

These equations give the number of evaluations required to achieve the threshold as a function of the number of samples taken per evaluation, but they do not indicate the total effort required to achieve the threshold. The total time required for the optimization procedure includes the time for the n samples taken at each evaluation and the overhead incurred by the GA for each evaluation. Taking these factors into consideration we arrive at two equations for the time necessary to achieve the threshold,

$$(7) \quad t_a = (\alpha_a + \beta_a n)(1244 + 2268\sigma/\sqrt{n})$$
$$(8) \quad t_b = (\alpha_b + \beta_b n)(18{,}020 + 9285\lambda/\sqrt{n})$$

where α is the GA overhead per evaluation and β is the time per sample. These equations allow us to determine the optimal value for n, i.e., the value which will minimize the time necessary to reach the desired threshold in this sample problem. It can be seen that for large n each expression for the time increases linearly with n. Thus, regardless of the relative size of the overhead, the optimal value of n is, not surprisingly, finite. As n approaches zero each expression approaches infinity, but the smallest possible value for n is one. The optimal value of n for either case can be found by finding the minimum of the appropriate expression subject to the restriction that n be an integer greater than zero. Further analysis requires some idea of the size of α/β. Since the results apply only to the particular example evaluation function $f(x,y,z)$ a detailed analysis is not worthwhile. We simply note that in the case in which α is negligible, the optimal value of n is 1, and as α increases the optimal value will increase. Thus, at least for small overhead the answer to the question concerning the relative importance of the number of evaluations versus the time required for a given evaluation is clear. The time required for a given evaluation is more important. The accuracy of the evaluation should be sacrificed in order to obtain more evaluations. Optimization proceeds more quickly with many rough evaluations than with few precise evaluations.

4.   An Experiment on Image Registration

The preceding simple example has the following special characteristics: (1) the function to be optimized is simple; (2) $r(x)$ has a normal distribution; (3) the standard deviation of $r(x)$ is a known function. These characteristics make it possible to do simple experiments which are easy to analyze. In more general problems these characteristics are not guaranteed, but they are not necessary to insure the efficacy of the statistical approach. To demonstrate the method for practical problems, we describe here our approach to a problem which has none of these characteristics. The problem is found in the registration of digital images. The functions which are optimized in image registration are measures of the difference between two images of a scene, in our case X-ray images of an area of a human neck, which have been acquired at different times. The images differ because of motion which has taken place between the two acquisition times, because of the injection of dye into the arteries, and because of noise in the image acquisition process. The registration of such images is necessary for the success of the process known as *digital subtraction angiography* in which an image of the interior of an artery is produced by subtracting a pre-injection image from a post-injection image. The details of the process and the registration technique can be found in [7]. By performing a geometrical transformation which warps one image relative to the other it is possible to improve the registration of the images so that the difference which is due to motion is reduced. The function parameters specify the transformation, and it is the goal of the genetic algorithms to find the parameter values which minimize the image difference.

The general problem of image registration is important in such diverse fields as aerial photography [8,16,17] and medical imaging [1,7,12,14,18]. General introductions to the field of image registration and extensive bibliographies may be found in [3,9,11]. An image comparison technique based on random sampling, different from the method used here, is described in [2]. The class of transformations which we consider includes elastic motion as well as rotation and translation.

The transformations which are employed here are illustrated in Figure 1. Two images are selected and a square subimage, the region of interest, is specified as image one -- im1. A geometrically transformed version of that image is to be compared to a second image -- im2. The transformation is specified by means of four vectors -- d1, d2, d3, and d4 -- which specify the motion of the four corners of im1. The transformed image is called im3. The motion of intermediate points is determined by means of bilinear interpolation from the corner points. The magnitudes of the horizontal and vertical components of the d vectors are limited to be less than one-fourth of the width of the subimage to avoid the possibility of folding [6]. (More complicated warpings will require additional

vectors.)

The images are represented digitally as square arrays of numbers representing an approximate map of image intensity. Each such intensity is called a *pixel*. The image difference is defined to be the mean absolute difference between the pixels at corresponding positions in im2 and im3. The exact mean can be determined by measuring the absolute difference at each pixel position; an estimate of the mean may be obtained by sampling randomly from the population of absolute pixel differences. The effort required to estimate the mean is approximately proportional to the number of samples taken; so, once again, the question arises as to the relative importance of number of evaluations used in the GA versus the time required per evaluation.

In general, the distribution of pixel differences for a given image transformation is not normal. Its shape will, in fact, depend in an unknown way on the geometrical transformation parameters, and consequently the standard deviation will change in an unknown way. Thus, while the experiments on $f(x,y,z)$ suggest that better results will be realized if less exact evaluations are made it is not clear how the level of accuracy should be set. We note that in the analysis of the experiments on $f(x,y,z)$ fixing the number of samples, n, has the effect of fixing, either σ_s or $\lambda_s = \sigma_s / f(x,y,z)$, given the assumed forms of σ. In the image registration case and in the general case, however, fixing n fixes neither of these quantities, since the σ's behavior cannot in general be expected to be so simple. We could, however, fix either of these quantities approximately by estimating σ using Equation (2) as samples are taken during an evaluation and continuing the sampling until n is large enough such that the estimate of σ_s obtained from Equation (1) is reduced to the desired value. Thus, the results from the previous experiments suggest three experiments on image registration -- (1) try to determine an optimal fixed n; (2) try to determine an optimal fixed σ_s; (3) try to determine an optimal fixed λ_s. We have implemented the first idea and a variation of the third idea. The variation is motivated by noting from statistical sampling theory that by fixing λ_s we are equivalently fixing our confidence in the accuracy of the sample mean as representative the actual mean. If, for example, we require that the sample mean be within $(100p)\%$ of the actual mean with 95% confidence, we should sample until we determine that λ_s is less than or equal to $p/1.96$ [19]. If we can fix only an estimate of λ_s, as in the general case, then the $(100p)\%$ accuracy at 95% confidence level requires that the estimate of λ_s be less than or equal to $p/t_{.05}(n)$. Here $t_\gamma(n)$ is student's t at a confidence level of $100(1-\gamma)\%$ and a sample size of n [4]. This *t-test* is exact only if the distribution of the sample mean is normal. In order to assure that the sample mean is approximately normal the sample size, n, should be at least 10 [4]. Our variation on fixing λ_s is to pick a confidence level of 95% (an arbitrary choice) and then fix p, subject to $n \geq 10$ to determine an optimal p.

The experiments to determine an optimal value of n and p for image registration and in the general case differ from those described for $f(x,y,z)$ above in two ways. First, because so little is known about the distributions in the general case, actual sampling is necessary. Second, because so little is known about the mean which is to be optimized (minimized) it is difficult to determine in the general case whether a threshold has been reached, and therefore the criterion for halting must be different. We have considered two alternative halting criteria: (1) determining an exact mean, or a highly accurate estimate of the mean, of the structure whose estimate is the best at each generation, halting when that value reaches a threshold, and using as a measure of performance the total number of samples taken; (2) halting after a fixed number of samples have been taken and using as the measure of performance the exact evaluation of the structure whose estimate is the best at the last generation. The first alternative suffers from the disadvantage that the additional evaluation at each generation is expensive and tends to offset the savings gained through approximate evaluation. The severity of the disadvantage is,

on the other hand, diminished as the size of the generation is increased. Therefore this method suggests a new consideration in setting the number of structures per generation. We choose in this work to avoid the question of the optimal number of structures by choosing the simpler alternative, (2).

The results of our experiments on image registration are shown in Figure 3. The Figure shows data resulting from 10 runs at each setting. The subimage im1 is 100 by 100 pixels, giving a sample space of size 10,000. The motion of the corners is limited to 8 pixels in the x and y directions. In each case the GA is halted after the generation during which the total number of samples taken exceed 200,000. The parameters for the transformation comprise the x and y components of the four d vectors. The range for each of these eight components is [-8.0, +8.0] digitized to eight bit accuracy. The GA parameters are set to optimize offline performance, as suggested by [10]: population size 80, crossover rate 0.45, mutation rate 0.10.

In Figure 3a each GA takes a fixed number of samples per evaluation. It can be seen from the Figure that the optimal sample size is approximately 10 samples per evaluation. Apparently, taking one sample per evaluation does not give the GA sufficient information to carry out an efficient search. The fact that performance deteriorates when we take fewer than 10 samples may indicate that the underlying distribution of pixel difference is not in general normal, and so this application does not correspond to the ideal experiments described in section 3.

In Figure 3b the estimated accuracy interval, based on the t-test, is fixed subject to the restriction that the sample size be at least 10. (Note that in Figure 3b, a 10% accuracy interval means that we are 95% confident that the sample mean is within 10% of the true mean.) These experiments suggest that the optimal accuracy interval at 95% confidence is nearly 100%, which corresponds to taking on the average 10 samples per evaluation. Given that the performance level is nearly identical whether we take exactly 10 samples per evaluation or we take on the average 10 samples, the first approach is preferable, since it does not require the calculation of the t-test for each sample.

It should be pointed out that, as in the experiment on $f(x,y,z)$, the GA overhead is ignored here. If the overhead were included, the optimal sample size would be somewhat larger. In any case, it is clear that a substantial advantage is obtained in statistical evaluation by reducing sampling sizes and accuracies, at least for this case of image registration.

5. Conclusions

GA's search by allocating trials to hyperplanes based on an estimate of the relative performance of the hyperplanes. One result of this approach is that the individual structures representing the hyperplanes need not be evaluated exactly. This observation makes GA's applicable to problems in which evaluation of candidate solutions can only be performed through Monte Carlo techniques. The present work suggests that in some cases the overall efficiency of GA's may be improved by reducing the time spent on individual evaluations and increasing the number of generations performed.

This works suggests some topics which deserve deeper study. First, the GA incurs some overhead in performing operations such as selection, crossover, and mutation. If the GA runs for many more generations as a result of performing quicker evaluations, this overhead may offset the time savings. Future studies should account for this overhead in identifying the optimal time to be spent on each evaluation. Second, it would be interesting to see how using approximate evaluations effects the usual kinds of performance metrics, such as online and offline performance. Finally, additional theoretical work in this area work be helpful, since experimental results concerning, say, the optimal sample size can be expected to be highly application dependent.

References

1. D. C. Barber, "Automatic Alignment of Radionuclide Images," *Phys. Med. Biol.* Vol. *27*(3), pp.387-96 (1982).

2. Daniel I. Barnea and Harvey F. Silverman, "A Class of Algorithms for Fast Digital Image Registration," *IEEE Trans. Comp.* Vol. *21*(2), pp.179-86 (Feb. 1972).

3. Chaim Broit, *Optimal Registration of Deformed Images*, Ph. D. thesis, Computer and Info. Sci., Univ. of Pennsylvania (1981).

4. Chapman and Schaufele, *Elementary Probability Models and Statistical Inference*, Xerox College Publ. Co., Waltham, MA (1970).

5. K. A. DeJong, *Analysis of the behavior of a class of genetic adaptive systems*, Ph. D. Thesis, Dept. Computer and Communication Sciences, Univ. of Michigan (1975).

6. J. Michael Fitzpatrick and Michael R. Leuze, "A class of injective two dimensional transformations," to be published.

7. J. M. Fitzpatrick, J. J. Grefenstette, and D. Van Gucht, "Image registration by genetic search," *Proceedings of IEEE Southeastcon '84*, pp.460-464 (April 1984).

8. Werner Frei, T. Shibata, and C. C. Chen, "Fast Matching of Non-stationary Images with False Fix Protection," *Proc. 5th Intl. Conf. Patt. Recog.* Vol. *1*, pp.208-12, IEEE Computer Society (Dec. 1-4, 1980).

9. Ardesir Goshtasby, *A Symbolically-assisted Approach to Digital Image Registration with Application in Computer Vision*, Ph. D. thesis, Computer Science, Michigan State Univ. (1983).

10. J. J. Grefenstette, "Optimization of control parameters for genetic algorithms", to appear in *IEEE Trans. Systems, Man, and Cybernetics* (1985).

11. Ernest L. Hall, *Computer Image Processing and Recognition Academic Press*, Inc., New York (1979).

12. K. H. Hohne and M. Bohm, "The Processing and Analysis of Radiographic Image Sequences," *Proc. 6th Intnl. Conf. Patt. Recog. Vol. 2*, pp.884-897, IEEE Computer Society Press (Oct. 19-22, 1982).

13. F. James, "Monte Carlo theory and practice," *Rep. Prog. Phys. Vol. 43*, p.73 (1980).

14. J. H. Kinsey and B. D. Vannellü, "Applic. of Digit. Image Change Detection to Diagn. and Follow-up of Cancer Involving the Lungs," *Proc. Soc. Photo-optical Instrum. Eng. Vol. 70*, pp.99-112, Society of Photo-optical Instr. Eng. (1975).

15. B. Lautrup, "Monte Carlo methods in theoretical high-energy physics," *Comm. ACM Vol. 28*, p.358 (April 1985).

16. James J. Little, "Automatic Registration of Landsat MSS Images to Digital Elevation Models," *Proc. Workshop Computer Vision: Representation and Control*, pp.178-84 IEEE Computer Science Press (Aug. 23-25, 1982).

17. Gerard G. Medioni, "Matching Regions in Aerial Images," *Proc. Comp. Vision and Patt. Recog.*, pp.364-65, IEEE Computer Society Press (June 19-23, 1983).

18. Michael J. Potel and David E. Gustafson, "Motion Correction for Digital Subtraction Angiography," *IEEE Proc. 5th An. Conf. Eng. in Med. Biol. Soc.*, pp.166-9 (Sept. 1983).

19. Murray R. Spiegel, *Theory and Problems of Probability and Statistics*, McGraw-Hill, New York (1975).

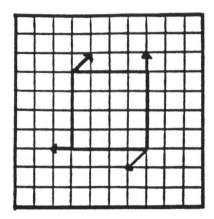

Figure 1a.

Subimage im1 is represented by the smaller inner square. The arrows represent the four d-vectors.

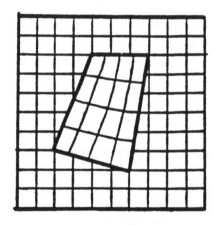

Figure 1b.

im2 is the larger image. im3. is the inner image formed by transforming im1 according to the d-vectors shown in Fig. 1a.

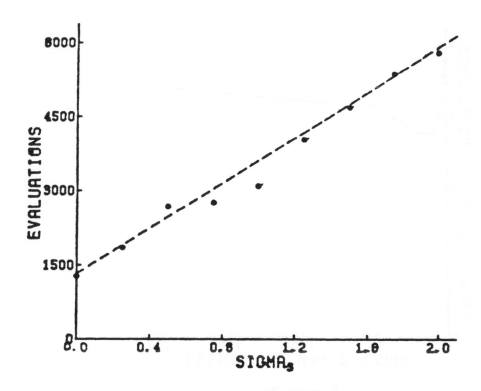

Figure 2a.
Evaluations Until Threshold vs. Absolute Error.

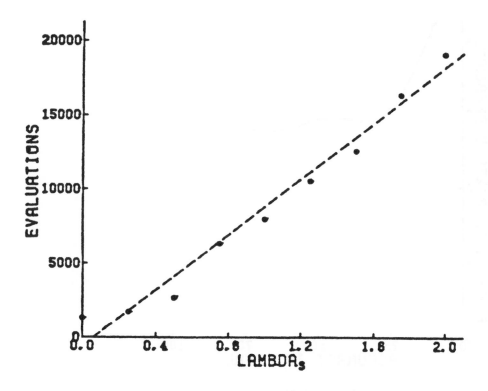

Figure 2b.
Evaluations Until Threshold vs. Relative Error.

119

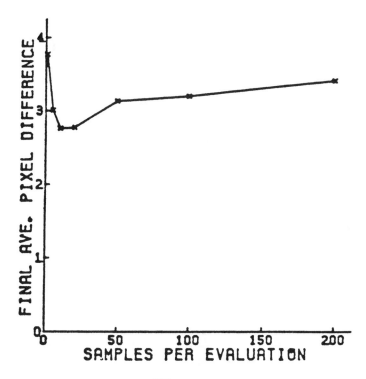

Figure 3a.
Performance vs. Fixed Sample Size.

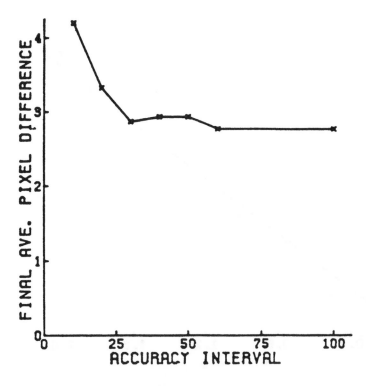

Figure 3b.
Performance vs. Accuracy Interval.

A connectionist algorithm for genetic search[1]

David H. Ackley
Department of Computer Science
Carnegie-Mellon University
Pittsburgh, PA 15213

Abstract

An architecture for function maximization is proposed. The design is motivated by genetic principles, but connectionist considerations dominate the implementation. The standard genetic operators do not appear explicitly in the model, and the description of the model in genetic terms is somewhat intricate, but the implementation in a connectionist framework is quite compact. The learning algorithm manipulates the gene pool via a symmetric converge/diverge reinforcement operator. Preliminary simulation studies on illustrative functions suggest the model is at least comparable in performance to a conventional genetic algorithm.

1 Overview

A new implementation of a genetic algorithm is presented. The possibility for it was noted during work on learning evaluation functions for simple games [1] using a variation on a recently developed connectionist architecture called a Boltzmann Machine [2]. The present work abstracts away from game-playing and focuses on relationships between genetic algorithms and massively parallel, neuron-like architectures.

This work takes function maximization as the task. The system obtains information by supplying inputs to the function and receiving corresponding function values. By assumption, no additional information about the function is available. Finding the maximum of a complex function possessing an exponential number of possible inputs is a formidable problem under these conditions. No strategy short of enumerating all possible inputs can always find the maximum value. Any unchecked point might be higher than those already examined. Any practical algorithm can only make plausible guesses, based on small samples of the parameter space and assumptions about how to extrapolate them.

However, the function maximization problem avoids two further complexities faced by more general formulations. First, performing "associative learning" or "categorization" can be viewed as finding maxima in specified subspaces of the possible input space. Second, in the most general case, the function may change over time, spontaneously or in response to the system's behavior. There the entire history of the search may affect the current location of the maximum value.

Section 2 presents the model. For those familiar with genetic algorithms, highlights of Section 2 are

- Real-valued vectors are used as genotypes instead of bit vectors. Reproduction and crossover are continuous arithmetic processes, rather than discrete boolean processes.

---

[1] This research is supported by the System Development Foundation.

- The entire population is potentially involved in each crossover operation, and crossover is not limited to contiguous portions of genes.

- The reproductive potential of genotypes is not determined by comparison to the average fitness of the population, but by comparison to a threshold. Adjusting the threshold can induce rapid convergence or diverge an already converged population.

Section 3 describes simulation studies that have been performed. The model is tested on functions that are constructed to explore its behavior when faced with various hazards. First a simple convex function space is considered, then larger spaces with local maxima are tried.

Section 4 discusses the model with respect to the framework of reproductive plans and genetic operators developed in [10]. Possible implications for connectionist research are not extensively developed in this paper.

Section 5 concludes the paper.

2 Development

The goal of this research was to satisfy both genetic and connectionist constraints as harmoniously as possible. As it turned out, the standard genetic operators appear only implicitly, as parts of a good description of how the model behaves. On the other hand, the implementation of the model in connectionist terms is not particularly intuitive. After sketching a genetic algorithm, this section presents the model via a loose analogy to the political process of a democratic society. The section concludes by detailing the implementation of this "election" model and drawing links between the genetic, the political, and the connectionist descriptions.

2.1 Genetic algorithms. Genetic evolution as a computational technique was proposed and analyzed by Holland [10]. It has been elaborated and refined by a number of researchers, e.g. [3, 4] and applied in various domains, e.g. [13, 6]. In its broadest formulations it is a very general theory; the following description in terms of function maximization is only one of many possible incarnations.

Genetic search can be used to optimize a function over a discrete parameter space, typically the corners of an n dimensional hypercube, so that any point in the parameter space can be represented as an n bit vector. The technique manipulates a set of such vectors to record information gained about the function. The pool of bit vectors is called the *population*, an individual bit vector in the population is called a *genotype*, and the bit values at each position of a genotype are called *alleles*. The function value of a genotype is called the genotype's *fitness* or *figure of merit*.

There are two primary operations applied to the population by a genetic algorithm. *Reproduction* changes the contents of the population by adding copies of genotypes with above-average figures of merit. The population is held at a fixed size, so below-average genotypes are displaced in the process. No new genotypes are introduced, but changing the distribution this way causes the average fitness of the population to rise toward that of the most-fit existing genotype.

In addition to this "reproduction according to fitness," it is necessary to generate new, untested genotypes and add them to the population, else the population will simply converge on the best one it started with. *Crossover* is the primary means of generating plausible new genotypes for addition to the population. In a simple implementation of crossover, two genotypes are selected at random from the population. Since the population is weighted towards higher-valued genotypes, a random selection will be biased in the same way. The crossover operator takes some of the alleles from one of the "parents" and some from the other, and combines them to produce a complete genotype. This "offspring" is added to the population, displacing some other genotype according to various criteria, where it has the opportunity to flourish or perish depending on its fitness.

To perform a search for the maximum of a given function, the population is first initialized to random genotypes, then reproduction and crossover operations are iterated. Eventually some (hopefully maximal valued) genotype will spread throughout the population, and the population is said to have "converged." Once the population has converged to a single genotype, the reproduction and crossover operators no longer change the makeup of the population.

One technical issue is central to the development of the proposed model. In addition to reproduction and the crossover operator, most genetic algorithms include a "background" *mutation* operator as well. In a typical implementation, the mutation operator provides a chance for any allele to be changed to another randomly chosen value. Since reproduction and crossover only redistribute existing alleles, the mutation operator guarantees that every value in every position of a genotype always has a chance of occuring. If the mutation rate is too low, possibly critical alleles missing from the initial random distribution (or lost through displacement) will have only a small chance of getting even one copy (back) into the population. However, if the probability of a mutation is not low enough, information that the population has stored about the parameter space will be steadily lost to random noise. In either of these situations, the performance of the algorithm will suffer.

2.2 A democratic society metaphor. Envision the democratic political process as a gargantuan function maximization engine. The political leanings of the voting population constitute the system's store of information about maximizing the nebulous function of "good government." An election summarizes the contents of the store by computing simple sums across the entire population and using the totals to fill each position in the government. When the winners are known, voters informally express opinions about how well they think the elected government will fare. The bulk of the time between elections is spent estimating how well the government actually performs. By the next election, this evaluation process has altered the contents of the store: better times favor incumbents; worse times, challengers.

In society, the function being optimized is neither well-defined nor arbitrary, and the final evaluation of a government must be left to history, but in the abstract realm of function maximization the true value of a point supplied to any function can be determined in a single operation. The immediacy and accuracy of this feedback creates an opportunity for an explicit learning algorithm that would be difficult to formalize in a real democracy. Credit and blame can be assigned to the *voters* based on how well their opinions about the successive governments *predict* the results produced by the objective function. Voters that approved of a high-scoring government can be rewarded by giving them more votes, so their preferences become a bit more influential in the subsequent election. Voters in such circumstances tend to favor the status quo. Voters whose preferences cause them to

123

approve of a low-scoring government lose voting power, and become a bit more willing to take a chance on something new. The proposed model is built around such an approach to learning.

An iteration of the algorithm consists of three phases which will be called "election," "reaction," and "outcome." The function maximization society is run by an n member "government" corresponding to the n dimensions of the function being maximized. In each election all n "government positions" are contested. There are two political parties, "*Plus*" and "*Minus*." A genotype represents a voter's current party preferences, recording a signed, real-valued number of votes for each of the positions. Which party wins a position depends on the net vote total for that position. A government represents a point in the parameter space, with *Plus* signifying a 1 and *Minus* signifying a 0.

After an election is concluded, each voter chooses a reaction to the new government: "satisfied," "dissatisfied," or "apathetic." The complete state of a voter includes the weights of its genotype plus its reaction. In general, voters whose genotypes match well with the government—i.e., most (or the most strongly weighted) of the positions have the same signs as the genotype weights—will be satisfied and therefore share in the credit or blame for the government's performance. Voters that got about half of their choices are likely to be apathetic, and therefore are unaffected by any consequent reward or punishment. Voters that got few of their choices are likely to be dissatisfied with the election results. Dissatisfied voters share in the fate of the government, but with credit and blame reversed in a particular way discussed below. Satisfied and dissatisfied voters are also referred to as *active*, and apathetic voters are also referred to as *inactive*.

In the outcome phase, the performance of the government is tested by supplying the corresponding point to the objective function and obtaining a function value. This value is compared to the recent history of function values produced by previously elected governments to obtain a reinforcement signal. A positive result indicates a point scoring better than usual and vice-versa. The reinforcement signal is used to adjust the preferences of the active voters. Positive reinforcement makes the reactions of the population more stable, and negative reinforcement makes them more likely to change. Finally, the newly obtained function value is incorporated into the history of function values, and an iteration is complete.

Two points are worth making before considering the actual implementation. The first point is that there is noise incorporated into both the election and the reaction processes. If the sum of the vote for a given position is a landslide, the result will essentially always be as expected, but as the vote total gets closer to zero the probability rises that the winner of the position will not actually be the party that got the most votes. There are no ties or runoff elections; if the sum of the vote for a position totals to exactly zero the winner is chosen completely at random. Voter reactions are also stochastic, based on the net degree of match over mismatch between each genotype and the elected point. Although real election systems try to ensure that the winner got the most votes, in the proposed model this nondeterminism serves the crucial function of introducing mutation. Moreover, unlike the constant-probability mutation operator mentioned in the previous section, it is *data dependent*. Mutation is very likely in those positions where no consensus arises from the population, but it will almost never upset a clear favorite.

The second point is that only the currently active voters participate in the election.

124

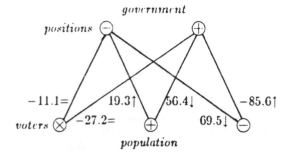

Point tested: $I = 01$
Point value: $v = 87.7$
Expectation level: $\theta = 113.8$
Reinforcement: $r = -0.3462$

Figure 1. A very small instance of the model. There are two positions and three voters. The state of the system is shown in the middle of an outcome phase, just before apportionment of credit. The election results are shown in the top row of circles; the ensuing voter reactions are shown in the bottom row. The first voter is apathetic. The second voter is satisfied, even though it didn't get its preference on the first position. The third voter is dissatisfied with the election results. On the right are the other data maintained by the model. The elected government corresponds to the binary vector 01, which has been passed to the objective function which returned the value 87.7. This is less than expected so the reinforcement signal is negative. The symbols next to the link weights indicate whether the weights will increase ($\uparrow$), decrease ($\downarrow$), or remain the same (=) when the reinforcement signal is applied. Negative reinforcement rewards inconsistency; in this example, only the first weight of the second voter increases in magnitude.

Satisfied voters vote in the manner described above. Dissatisfied voters vote in a sign-reversed manner: positive weights vote for *Minus* and negative weights vote for *Plus*. Apathetic voters do not vote at all, but they react to each election and may become active. Section 4 discusses a genetic interpretation of this strategy.

2.3 A connectionist implementation. The ever-increasing demand for computational power and the continuing desire to understand the human brain has encouraged research into massively parallel computational architectures that resemble the physiological picture of the brain more closely than does the standard Von Neumann model. The basic assumption of the connectionist approach (see, e.g., [5] or [7]), is that computation can be accomplished collectively by large numbers of very simple processing units that contain very little storage. The bulk of the memory of the system is located in communication links between the units, usually in the form of one or a few scalar values per link that control the link's properties. In terms of individual units and links, the Perceptron [12] typifies the kinds of hardware considered: a unit is simple linear threshold device, adopting one of two numeric output states based on a comparison between the sum of its input links and its threshold; a link connects two units and contains a scalar variable that is multiplied by the link input to produce the link output.

In terms of problem formulations, network organizations, and learning algorithms, connectionist research has moved in many directions from the Perceptron; the proposed model uses assumptions most closely related to those employed in [1, 2, 9, 11]. There is not space to explicitly motivate all of the decision designs of the implementation, but analogies to the political and genetic descriptions are discussed as they arise. Figure 1 sketches an instance of the model and defines terminology.

The basic processing element of the model is called a *unit*. Each unit i has a ternary

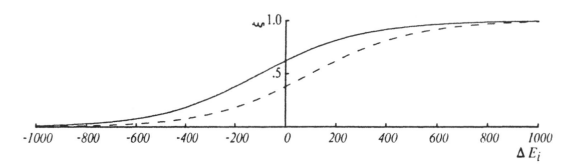

Figure 2. A phase diagram of s_i as a function of ΔE_i and ξ. The boundary curves are specified by Eq. (2), plotted at $T = 200$ and $\alpha = 100$. $(\Delta E_i, \xi)$ points falling above the solid line generate $s_i = -1$, points below the dotted line generate $s_i = 1$, and points between the lines generate $s_i = 0$.

state variable $s_i \in \{+1, 0, -1\}$. Units communicate their current states to other units via *links*. A link between two units i and j has a real-valued weight w_{ij}. All links between units are bidirectional and have the same weight in both directions, i.e. $w_{ij} = w_{ji}$.

In the political analogy, groups of units represent both the government positions and the voters. In the former case, s_i represents the winner of position i, with $s_i = 1 \rightarrow$ *Plus* and $s_i = -1 \rightarrow$ *Minus*. Parameters are set so that $s_i = 0$ cannot occur for the position units. In the latter case, s_i represents the reaction of voter i, with $s_i = 1 \rightarrow$ "satisfied," $s_i = 0 \rightarrow$ "apathetic," and $s_i = -1 \rightarrow$ "dissatisfied."

A unit simply retains its current state until it is *probed*, at which time it checks the states of the units it is connected to and the weights on those links and applies a probabilistic decision rule to select a state. The quantity that sums up the current context of a unit i is called "ΔE_i" and is defined as

$$\Delta E_i = 2 \sum_j s_j w_{ij} \, . \tag{1}$$

where j ranges over all the units in the network and $w_{ij} = 0$ if units i and j are not connected. Given ΔE_i and a uniform random variable $0 \le \xi < 1$, the decision rule is

$$s_i = \begin{cases} +1 & \text{if } \xi \ge \dfrac{1}{1 + e^{-(\Delta E_i + \alpha)/T}} \\[2mm] -1 & \text{if } \xi < \dfrac{1}{1 + e^{-(\Delta E_i - \alpha)/T}} \\[2mm] 0 & \text{otherwise.} \end{cases} \tag{2}$$

The boundaries between the unit states are plotted in Figure 2. The size of the model parameter $T > 0$—the "temperature"—determines how sharply the boundaries slope as ΔE_i moves away from zero; it controls how "noisy" the system is. The model parameter $\alpha \ge 0$ controls the width of the "apathy window" when the voter units are probed.

In the political analogy, the election and reaction processes are both implemented by the probe operation. An election is performed by probing each of the position units once. Since position units connect only to voter units the ordering of the probes is irrelevant, and the contests for each position can happen in parallel. When applied to a position

126

unit i, the summation in Eq. (1) totes up the effective vote count for the position. If a voter unit j is apathetic, then $s_j = 0$ and w_{ij} does not affect the total for the position, otherwise either w_{ij} or $-w_{ij}$ is included in the total depending on whether the voter is satisfied or dissatisfied. The winner of the position is then determined by Eq. (2), applied with $\alpha = 0$. As ΔE_i becomes more positive, the likelihood of *Plus* winning the contest increases, and vice-versa. If one takes the limit as $T \to 0$, Eq. (2) approaches a step function corresponding to a deterministic election based only on the sign of ΔE_i.

The voter reaction is assessed symmetrically, by probing each of the voter units once. When applied to a voter unit i, the summation in Eq. (1) produces a net match score between an elected government and the voter's preferences. The match score for the voter increases when the state of position j has the same sign as w_{ij} and decreases when the signs differ. The voter's reaction is then determined by Eq. (2), with α set as a model parameter. A large positive ΔE_i indicates a particularly good match between a government and a voter, and generates a high probability that the voter will be satisfied and adopt $s_i = 1$; a large negative value indicates a particularly bad match and strongly suggests $s_i = -1$; and a near-zero value indicates an ambiguous situation and generates the largest probability of adopting $s_i = 0$. The assumption of bidirectional links with symmetric weights guarantees that a voter's behavior during elections and reactions will be consistent. If all of a voter's preferred candidates are elected, for example, then in the zero temperature limit the voter cannot be dissatisfied with the government.

In genetic terms, an election can be viewed as part of a generalized crossover operation. If we imagine one satisfied voter in an otherwise apathetic population, the outcome of a (sufficiently low temperature) election will be a direct expression of that voter's genotype: wherever the weight from the voter to a position is positive *Plus* will win and vice-versa. If two voters are satisfied, some mixture of their genotypes will be expressed by the position units, depending on the relative magnitudes of the weights to the positions where the voters disagree. This situation bears a close resemblance to the standard crossover operator. The difference is that standard crossover determines the winners of disputed positions by a random choice of crossover point, whereas the proposed model exploits accumulated performance data to bias each decision.[2] In the general case the crossover operation is hard to see explicitly, considering the effects of many satisfied voters, the dissatisfied vote, temperature, and the fact that the crossed-over genotype is not guaranteed admission to the population.

The next steps in the algorithm are straightforward. The states of the position units are translated into a binary vector I; the vector is passed to the objective function; a scalar value v is returned. The function value has no meaning in itself since the possible range of function values is unknown. A judgment must be made whether the value is "good" or "bad," assuming that whatever is deemed good will be made more probable in the future. The expectation level θ is used to produce the reinforcement signal

$$r = \frac{2}{1 + e^{(\theta - v)/T_r}} - 1. \tag{3}$$

---

[2] This statement is too strong if the model using standard crossover also uses inversion, since in that case the grouping induced by the crossover point does depend on the past performance of the model, as recorded by the inversion operator. Section 4 discusses inversion and crossover further.

> 0. *Initialization:* Given unknown function $\{v = f(I) | I \in 2^n, v \in \Re\}$. Select
>   model parameters. Create n position units and m voter units. Link each
>   position unit to each voter unit. Set all nm link weights $w_{ij} = 0$. Set all
> $n + m$ unit states $s_i = 0$. Set $\theta = 0$.
> 1. *Election:* Probe each position unit (Eqs. 1 and 2).
> 2. *Reaction:* Probe each voter unit.
> 3. *Outcome:*
>   3.1. *Fitness test:* Compute $v = f(I)$.
>   3.2. *Discount expectations:* Compute r (Eq. 3).
>   3.3. *Apportion credit:* Update w_{ij} (Eq. 4).
>   3.4. *Adjust expectations:* Update θ (Eq. 5).
> 4. *Iterate:* Go to step 1.
>
>
> $m > 0$ Size of population; number of voters.
> $T > 0$ Temperature of unit decisions.
> $\alpha \geq 0$ Apathy window for voter reactions.
> $k > 0$ Payoff rate.
> $T_r > 0$ "Temperature" of reinforcement scaling.
> $0 \leq \rho < 1$ Time constant for function averaging.
> $\delta \geq 0$ Excess expectation.

Figure 3. Algorithm summary and list of model parameters.

This employs the same basic sigmoid function used in the unit decision rule, but r is bounded by ± 1 and is used as an analog value rather than a probability. The model parameter T_r scales the sensitivity around $\theta = v$.[3] r is used to update the weights

$$w_{ij,t+1} = w_{ij,t} + krs_is_j \qquad (4)$$

where $k > 0$ is the payoff rate. The change to each link weight depends on the product s_is_j. If the voter unit is apathetic the weight does not change, otherwise either kr or $-kr$ is added to the weight, depending if the voter and position units are in the same or different states.

If r is positive, the net effect of this is that the ΔE of satisfied units becomes more positive and the ΔE of dissatisfied units becomes more negative, i.e., each active unit becomes somewhat less likely to change state when probed. Consistency is encouraged; the incumbents are more likely to be reelected, the voters are less likely to change their reactions. When r is negative the reverse happens. Inconsistency is encouraged; victory margins erode, voter reaction becomes more capricious. An updating of weights with positive r is called "converging on a genotype," with negative r, "diverging from a genotype."

In genetic terms, the weight modification procedure both implements reproduction and completes the implementation of the crossover operator. Only the crossed-over genotype as expressed in the position units is eligible for reproduction, and then only if $r > 0$. Otherwise the network diverges, and that genotype decreases its "degree of existence" in the population. It is displaced, by some amount, but it is not replaced with other

---

[3] The precise form of Eq. (3) does not appear essential to the model. Several variations all searched effectively, though they displayed different detailed behaviors.

members of the population—the total "voting power" of the population declines a bit instead. Intuitively speaking, the space vacated by a diverged genotype is filled with noise.

The final implementation issue is the computation of the expectation level. A number of workable ways to manipulate θ have been tried, but the simulations in the next section all use a simple backward-averaging procedure

$$\theta_{t+1} = \rho\theta_t + (1 - \rho)(v + \delta) \tag{5}$$

where $0 \leq \rho < 1$ is the "retention rate" governing how quickly θ responds to changes in v. Just allowing θ to track v is inadequate, however, for if the network completely converged there would be no pressure to continue searching for a better value. A positive value for the model parameter δ avoids this complacency and ensures that a converged network will receive more divergence than convergence, and eventually destabilize.

Figure 3 summarizes the algorithm and lists the seven model parameters.

3 Behavior

This section describes preliminary simulations of the election model. Most of the objective functions considered here were explored during the design of the model, rather than being chosen as independent tests after the design stabilized. The functions were created to embody interesting characteristics of search spaces in general.

All of the simulations described in this paper use the following settings for the model parameters

$$m = 50 \qquad T = 10n \qquad \alpha = 5n \qquad k = 2.0$$
$$T_r = 1.0 \qquad \rho = 0.75 \qquad \delta = 4.0$$

Note that the temperature and the apathy are proportional to the dimensionality of the given parameter space. For convenience, these are called the "standard" settings, but significantly faster searching on a function of interest can be produced by fine-tuning the parameters. The standard settings were chosen because they produce moderately fast performances across the four selected functions, each tested at four dimensionalities.

The simulations count the average number of function evaluations before the model evaluates the global maximum. Two other algorithms were implemented for comparison. The first was the following hillclimbing algorithm

1. Select a point at random and evaluate it.

2. Evaluate all adjacent points. If no points are higher than the selected point, go to step 1. Otherwise select the highest adjacent point, and repeat this step.

Iterated hillclimbing is a simple-minded algorithm that requires very little memory. Its performance provides only a weak bound on the complexity of a parameter space. The second algorithm was a basic version of Holland's R1 reproductive plan [10], using only simple crossover and mutation. Considering the lack of sophisticated operators in the implementation, and the author's inexperience at tuning its parameters, the performance of the R1 implementation should be taken only as an upper bound on the achievable performance of a simple genetic algorithm.[4]

---

[4]   The R1 model parameter values were selected after a short period of trial and error on the test

3.1 A convex space. Consider the following trivial function: *Score 10 points for each 1 bit. Return the sum.* The global maximum equals $10n$ and occurs when all bits are turned on. This "one max" function was tested because it can be searched optimally by hillclimbing, and the generality of a genetic search is unnecessary. Figure 4 tabulates the simulation results for $n = 8, 12, 16, 20$. As expected, the hillclimbing algorithm found the maximum more quickly than did the model, but it is encouraging that on all but the smallest case the election model comes within a factor of two of hillclimbing's efficiency on this convex space. Observations made during the simulations suggest that the relatively poorer performance of R1 arose primarily from the occasional loss of one or more critical alleles, producing the occasional very long run. Although increasing the mutation rate reduced the probability of such anomalies, it produced a costly rise in the length of typical runs.

One max				
n	8	12	16	20
Method	*Evaluations performed**			
Hillclimb	31	82	128	198
Election	73	117	187	302
Holland R1	195	674	1807	4161

* Rounded averages over 25 runs.

Figure 4. Comparative simulation results on the "one max" function. In all simulations, the performance measure is the number of objective function evaluations performed before the global maximum is evaluated.

3.2 A local maximum. Convex function spaces are very easy to search, but spaces of interest most often have local maxima, or "false peaks." Consider this "two max" function: *Score 10 points for each 1 bit, score -8 points for each 0 bit, and return the absolute value of the sum.* This function has the global maximum when the input is all 1's, but it also has a local maximum when the input is all 0's. Figure 5 summarizes the simulation results. With this function, a simple hillclimber may get stuck on the local maximum, so multiple starting points may be required.

Two max				
n	8	12	16	20
Method	*Evaluations performed**			
Hillclimb	37	97	186	230
Election	83	152	194	269
Holland R1	113	340	794	1622

* Rounded averages over 25 runs.

Figure 5. Comparative simulation results on the "two max" function.

---

functions. Using the notation defined in [10], the values were $M = 50$, $P_C = 1$, $P_I = 0$, $^1P_M = 0.5$, and $c_t = (1/t)^{0.25+2/n}$, where n is the dimensionality of the objective function. Constant offsets were added to the functions where necessary to ensure non-negative function values.

Nonetheless, on this function also the hillclimber outperforms the model, although only by a narrow margin on the larger cases. The mere existence of a local maximum does not imply that a space will be hard to search by iterated hillclimbing. The regions surrounding the two maxima of the function have a constant slope of 18 points per step toward the nearer maximum. The slopes have the same magnitude, so the higher peak must be wider at its base. With every random starting point, the hillclimber is odds on to start in the "collecting area" of the higher peak, so it continues to perform well.

3.3 Fine-grained local maxima. Consider the following "porcupine" function: *Score 10 points for each 1 bit and compute the total. If the number of 1 bits is odd, subtract 15 points from the total. Return the total.* Every point that has an even number of 1 bits is a porcupine "quill," surrounded on all sides by the porcupine's "back"—lower valued points with odd numbers of 1 bits. As the total number of 1 bits grows, the back slopes upward; the task is to single out the quill extending above the highest point on the back.

Porcupine				
n	8	12	16	20
Method	Evaluations performed*			
Hillclimb	145	2474	41973	–
Election	160	211	241	495
Holland R1	163	739	1296	3771

* Rounded averages over 25 runs.

Figure 6. Comparative simulation results on the "porcupine" function.

Unlike the first two functions, the porcupine function presents a tremendously rugged landscape when one is forced to navigate it by changing one bit at time. Not surprisingly, hillclimbing fails spectacularly here. Figure 6 displays the results. The landscape acts like flypaper, trapping the hillclimber after at most one move, and the resulting long simulation times reflect the exponential time needed to randomly guess a starting point within a bit of the global maximum. (The hillclimber was not run with $n = 20$ for that reason.) On the other hand, the election model gains less than a factor of two over its performance on the one max function. The strong global property of the space—the more 1's the better, other things being equal—is detected and exploited by both genetic algorithms.[5]

Although the porcupine function reduced hillclimbing to random combinatoric search, in a sense it cheated to do so, by exploiting the hillclimber's extremely myopic view of possible places to move. A hillclimber that considered changing two bits at a time could proceed directly to the highest quill. But increasing the working range of a hillclimber exacts its price in added function evaluations per move, and can be foiled anyway by using fewer, wider quills (e.g., subtract 25 points unless the number of ones is a multiple of

---

[5] The concept of parity, which determines whether one lands on quill or back, is not detected or exploited. All three algorithms continue to try many odd parity points during the search. The general notion of parity, independent of any particular pattern of bits, cannot be represented in such simple models; the import of this demonstration is that the genetic models can make good progress even when there are aspects of the objective function that, from their point of view, are fundamentally unaccountable.

three.) Higher peaks may always be just "over the horizon" of an algorithm that searches fixed distances outward from a single point.

3.4 Broad plateaus. The porcupine function was full of local maxima, but they were all very small and narrow. A rather different sort of problem occurs when there are large regions of the space in which all points have the same value, offering no uphill direction. Consider the following "plateaus" function: *Divide the bits into four equal-sized groups. For each group, if all the bits are 1 score 50 points, if all the bits are 0 score -50 points, and otherwise score 0 points. Return the sum of the scores for the four groups.* In a group, any pattern that includes both zeros and ones is on a plateau. Between the groups the bits are completely independent of each other; within a group only the *combined* states of all the units has any predictive power. When $n = 8$ there are only two bits in a group and the function space is convex, because the sequence $00 \rightarrow \{01, 10\} \rightarrow 11$ is strictly uphill. However, since each group grows larger as n increases, this function rapidly becomes very non-linear and difficult to maximize.

Plateaus				
n	8	12	16	20
Method	*Evaluations performed**			
Hillclimb	34	414	2224	13404
Election	146	392	758	2364
Holland R1	228	697	2223	8197

* Rounded averages over 25 runs.

Figure 7. Comparative simulation results on the "plateaus" function.

4 Discussion

The proposed model was developed only recently, and has it has not yet been analyzed or tested extensively. Although it would be premature to interpret the model and simulations in a very broad scope, a few interesting consequences have been uncovered already. This section touches on a number of relationships between the election model and the analytic structure of schemata and generalized genetic operators developed by Holland in *Adaptation in Natural and Artificial Systems* (ANAS) [10].

Given a population, computational effects related to simple crossover can be achieved in many ways. For example, disputed positions could be resolved by random choices between the parents, or by appealing to a third genotype as a tie-breaker. Like simple crossover, both of these implementations perform the basic task of generating new points that instantiate many of the same schemata as the parents. An appropriate crossover mechanism interacts well with the other constraints of the model and the task domain. For example, the information represented by a DNA molecule is expressed linearly, so the sequential ordering of the alleles is critical. In these circumstances, the simple cut-and-swap crossover mechanism is an elegant solution, since it is cheap to implement and it preferentially promotes contiguous groups of co-adapted alleles.

In an unconstrained function optimization task, as little as possible should be presumed *a priori* about how the external function will interpret the alleles. In these circumstances, the sequential bias of the standard crossover mechanism is unwarranted. ANAS proposes an *inversion* operator to compensate for it. The inversion operator tags each allele with its position number in terms of the external function, so the ordering of the genotype can be permuted to bring co-adapted alleles closer together and therefore shelter them from simple crossover. However, if two chosen parents do not have their genotypes permuted in the same way, a simple crossover between them may not produce a complete set of alleles. ANAS offers two suggestions. If inversion is a rare event, sub-populations with matching permutations can develop, and crossover can be applied only within such groups. But then information about the linkages between alleles accumulates only slowly. Alternatively, one of the parents can be temporarily permuted to match the other parent in order to allow simple crossover to work, but then half of the accumulated linkage information is ignored at each crossover.

The proposed model does not use the ordering of the alleles to carry information. Linkage information is carried in the magnitudes of the genotype weights, in non-obvious ways involving all three phases and the assumption of symmetric weights. For example, the defining loci of a discovered critical schema are likely to be represented by relatively large weights on a genotype, since those weights will receive systematically higher net reinforcement than the non-critical links. Conversely, relatively large weights to a few positions cause the designated alleles to behave in a relatively tightly coupled fashion. In the election phase, large weights increase the chance that the alleles will be expressed simultaneously and receive reproduction opportunities. In the reaction phase, the same large weights increase the chance that the voter will be apathetic when the implied schema is *not* expressed, since the genotype's large weights will tend to cancel. Strongly coupled alleles will be disrupted more slowly over successive outcome phases.

Although it is not discussed in ANAS, subsequent research found it useful to include a "crowding factor" that affects how genotypes get selected for deletion to make room for a new offspring [4]. The idea is to prefer displacing genotypes that are similar to the new one, thus minimizing the loss of schemata. In the proposed model, note the interaction between the reaction phase and the outcome phase. Only active voters are affected by weight modification. Since voters tend to be satisfied or dissatisfied when they strongly match or mismatch the government, and dissatisfied voters invert the sign of the weight modifications, converging on a genotype preferentially displaces *similar* existing genotypes.

The representation of genotypes by real-valued vectors instead of bit vectors has widespread consequences. One major difference concerns the displacement of genotypes as a result of reproduction or crossover. When a bit vector is displaced from a conventional population, the information it contained is permanently lost. In contrast, the proposed reinforcement operator is an invertible function. Between a constant government and a voter, any sequence of positive and negative reinforcements has the same effect as their sum. Observations revealed that the election model exploits this property in an unanticipated and useful way. The happenstance election of a surprisingly good government often leads to a run of reelections and positive reinforcements, occasionally freezing the network solid for a few iterations, until the expectation level catches up. If one examines the signs of the genotype weights at such a point and interprets them as boolean variables, the population often looks nearly converged. But the expectation level soon exceeds any fixed value, and weaker negative reinforcements begin to cancel out the changes and to *regenerate* the pre-convergent diversity. During such times, the government positions with the smallest victory margins are the first to begin changing, which causes a period of stochastic local search in an expanding neighborhood around the convergence point. If further improvement is discovered, the network will frequently converge on it, but often the destabilization spreads until the government collapses entirely and a period of wide-ranging global search ensues. It may be that much of the election model's edge over the R1 algorithm on the strict maximization-time performance metric used in this paper arises from this tendency to hillclimb for a while in promising regions of the parameter space, without irrevocably converging the population.

5 Conclusion

The architectural assumptions of the model—the unit and link definitions, the decision rule, and the weight update rule—were first explored for reasons unrelated to genetic algorithms. The assumption of symmetric links between binary (± 1) threshold units was made by Hopfield [11] because he could prove such networks would spontaneously minimize a particular "energy" function that was easily modifiable by changing link weights. Hopfield used the modifiable "energy landscape" to implement an associative memory.

Hopfield's deterministic decision rule was recast into a stochastic form by Hinton & Sejnowski [8] because they could then employ mathematics from statistical mechanics to prove such a system would satisfy an asymptotic log-linear relationship between the probability of a state and the energy of the state. 0/1 binary units were used. They found a distributed learning algorithm that would provably hillclimb in a global statistical error measure. They used the system to learn probability distributions.

The weight update rule was investigated by the author because it provided a simple method of adjusting energies of states based on a reinforcement signal for a back-propagation credit assignment algorithm [1]. ± 1 binary units were used. The connectionist network was used as a modifiable evaluation function for a game-playing program. The system learned to beat simple but non-trivial opponents at tic-tac-toe. Observations made during simulations raised the possibility that genetic learning was occurring as the system evolved. In that work, the government corresponds to the game board, and a voter, in effect, specifies a sequence of moves and countermoves for an entire game. The model frequently played out variations that looked like crossed-over "hybrid strategies." The rapid spread through the units of a discovered winning strategy was suggestive of a reproduction process.

The research reported here focused on that possibility. The task was simplified to avoid problems caused by legal move constraints, opposing play, and delayed reinforcement. Given an appropriate problem statement, the basic election/reaction scheme seemed to be the simplest approach. Extending the unit state and decision rule to three values occurred to the author while developing the political analogy. In theory, apathy could be eliminated, because a unit with a near-zero ΔE would pick $+1$ or -1 randomly, so rewards and punishments irrelevant to that unit's genotype would cancel out in the long run. In practice, explicitly representing apathy improves the signal-to-noise ratio of the reinforcement signal with respect to the genotype. The unit is not forced to take a position and suffer the consequences when it looks like a "judgment call." The performance of the algorithm is generally faster and more consistent, but a percentage of the population is ignored at each election. For the large populations implied by massively parallel models, it appears to be an attractive space/time trade-off.

The connectionist model presented here has a much more sophisticated genetic description than was anticipated at the outset. Only reproduction, crossover and mutation were intentionally "designed into" the model. It was a surprise to discover that the model performed functions reminiscent of other genetic operators such as inversion and crowding factors. As an emergent property, the model displays both local hillclimbing and global genetic search, shifting between strategies at sensible times. More experience with the proposed model is needed, but a crossing-over of genetic and connectionist concepts appears to have produced a viable offspring.

References

[1] Ackley, D.H. Learning evaluation functions in stochastic parallel networks. Carnegie-Mellon University Department of Computer Science thesis proposal. Pittsburgh, PA: December 4, 1984.

[2] Ackley, D.H., Hinton, G.E., & Sejnowski, T.J. A learning algorithm for Boltzmann Machines. *Cognitive Science*, 1985, *9*(1), 147–169.

[3] Bethke, A.D. Genetic algorithms as function optimizers. University of Michigan Ph.D. Thesis, Ann Arbor, MI: 1981. .

[4] DeJong, K.A. Analysis of the behavior of a class of genetic algorithms. University of Michigan Ph.D. Thesis, Ann Arbor, MI: 1975.

[5] Feldman, J., (Ed.) Special issue: Connectionist models and their applications. *Cognitive Science*, 1985, *9*(1).

[6] Goldberg, D. Computer aided gas pipeline operation using genetic algorithms and rule learning. University of Michigan Ph.D. Thesis (Civil engineering), Ann Arbor, MI: 1983.

[7] Hinton, G.E. & Anderson, J.A. *Parallel Models of Associative Memory*. Hillsdale, NJ: Erlbaum, 1981.

[8] Hinton, G.E., & Sejnowski, T.J. Optimal perceptual inference. *Proceedings of the IEEE Computer Society Conference of Computer Vision and Pattern Recognition*. June 1983, Washington, DC, 448–453.

[9] Hinton, G.E., Sejnowski, T.J., & Ackley, D.H. *Boltzmann Machines: Constraint satisfaction networks that learn*. Technical report CMU-CS-84-119, Carnegie-Mellon University, May 1984.

[10] Holland, J.H. *Adaptation in Natural and Artificial Systems*. University of Michigan Press, 1975.

[11] Hopfield, J.J. Neural networks and physical systems with emergent collective computational abilities. *Proceedings of the National Academy of Sciences USA*, 1982, *79*, 2554–2558.

[12] Rosenblatt, F. *Principles of neurodynamics: Perceptrons and the theory of brain mechanisms*. Washington, DC: Spartan, 1961.

[13] Smith, S. A learning system based on genetic algorithms. University of Pittsburgh Ph.D. Thesis (Computer science). Pittsburgh, PA: 1980.

Job Shop Scheduling with Genetic Algorithms

Dr. Lawrence Davis

Bolt Beranek and Newman Inc.

1. INTRODUCTION

The job shop scheduling problem is hard to solve well, for reasons outlined by Mark Fox *et al*[1]. Their chief point is that realistic examples involve constraints that cannot be represented in a mathematical theory like linear programming. In ISIS, the system that Fox *et al* have built, the problem is attacked with the use of multiple levels of abstraction and progressive constraint relaxation within a frame-based representation system. ISIS is a deterministic program, however, and faced with a single scheduling problem it will produce a single result. Given the vast search space where such unruly problems reside, the chances of being trapped on an inferior local minimum are good for a deterministic program. In this paper, techniques are proposed for treating the problem non-deterministically, with genetic algorithms.

2. JOB SHOP SCHEDULING: THE PROBLEM

A job shop is an organization composed of a number of work stations capable of performing operations on objects. Job shops accept contracts to produce objects by putting them through series of operations, for a fee. They prosper when the sequence of operations required to fill their contracts can be performed at their work centers for less cost than the contracted amount, and they languish when this is not done. Scheduling the day-to-day workings of a job shop (specifying which work station is to perform which operations on which objects from which contracts) is critical in order to maximize profit, for poor scheduling may cause such problems as work stations standing idle, contract due dates not being met, or work of unacceptable quality being produced.

The scheduling problem is made more difficult by that fact that factors taken into account in one's schedule may change: machines break down, the work force may be unexpectedly diminished, supplies may be delayed, and so on. A job shop scheduling system must be able to generate schedules that fill the job shop's contracts, while keeping profit levels as high as practicable. The scheduler must also be able to react quickly to changes in the assumptions its schedules are based on.

In what follows, we shall consider a simple job shop scheduling problem, intended to be instructive rather than realistic, and show how genetic algorithms can be used to solve it.

3. SJS-A SIMPLIFIED JOB SHOP

SJS Enterprises makes widgets and blodgets by contract. There are six work stations in SJS. Centers 1 and 2 perform the grilling operation on the raw materials that are delivered to the shop. Centers 3 and 4 perform the filling operation on grilled objects, and centers 5 and 6 perform the final milling operation on filled objects. Widgets and blodgets go through these three stages when they are manufactured. Thus, the sequence of processes to turn raw materials into finished objects is this:

RAW MATERIALS - GRILLING - FILLING - MILLING - CUSTOMER.

SJS has collected a number of statistics about its operations. Due to differences in its machinery and personnel, the expected time for a work station to complete its operation on an object is as follows, in minutes:

WORK STATION	WIDGETS	BLODGETS
1	5	15
2	8	20
3	10	12
4	8	15
5	3	6
6	4	8

The cost of running each of the work stations at SJS is as follows, per hour:

WORK STATION	IDLE	ACTIVE
1	10	70
2	20	60
3	10	70
4	10	70
5	20	80
6	20	100

In addition, SJS has overhead costs of 100 units per hour.

Finally, it requires some time for a work station to change from making widgets to making blodgets (or vice versa). The change time for each station is:

WORK STATION	CHANGE TIME
1	30
2	10
3	20
4	20
5	9
6	18

4. A SCHEDULING PROBLEM

Suppose SJS is beginning production with two orders, one for 10 widgets and one for 10 blodgets. How should it be scheduled so as to maximize profits from the point at which operations begin, to the point at which both orders are filled? Let us consider three schedules that address this problem.

In schedule 1, individual work stations are assigned their own contracts. We notice that the production of blodgets takes longer than the production of widgets, and so we make widgets with centers 2, 4, and 6, and make blodgets with centers 1, 3, and 5. If the shop follows this schedule, the various work stations are occupied as follows:

STATION	CONTRACT	WORKING	WAITING	HRS-WORKED	COST
1	blodgets	0-150	30	3	210
2	widgets	0-80	40	2	140
3	blodgets	15-162	60	3	210
4	widgets	8-88	40	2	150
5	blodgets	27-168	120	3	240
6	widgets	16-92	80	2	220

In simulating the operation of the job shop under this plan, we note that some work stations spend a good deal of time waiting for objects to work on. Work stations 5 and 6, for example, spend from one to two hours waiting because they are faster than the centers that feed objects to them. It is possible to let them stand idle for the first hour of the day without delaying the filling of the orders, yielding a second schedule with cost 970, a 17 per cent reduction over the first schedule, achieved by giving these work stations an initial idle hour. A different way to cut down on the waiting time would be to leave work station 6 idle throughout the day, performing all operations with work station 5 during the second and third hours of the day. Work station 5 must start work on blodgets when it begins, switch to widgets later on and finish them, then switch back to making blodgets at the end. The cost of this schedule is 950, an 18.8 per cent reduction over the direct cost of the first schedule.

It is interesting to note that a deterministic system would be likely to try one or the other of the two optimizations on the first schedule, but not both. Each of these optimizations brings the situation to a local minimum in cost, and advance predictions of which such optimization will be best appear difficult to make.

5. AN AMENABLE REPRESENTATION OF THE PROBLEM

If we consider a schedule to be a literal specification of the activity of each work station, perhaps of the form "Work station w performs operation o on object x from time t1 to time t2," then one will be caught in a dilemma if one applies genetic techniques to this problem. Either one will attempt to use **CROSSOVER** operations or not. If so, their use will frequently change a legal schedule into an illegal one, since exchanging such statements between chromosomes will cause operations to be ordered for which the necessary previous operations have not been performed. As a result, one would acquire the benefits of **CROSSOVER** operations at the cost of spending a good deal of one's time in a space of illegal solutions to the problem. If one foregoes **CROSSOVER** operations, however, one loses the ability to accelerate the search process, the very feature of the genetic method that gives it its great power.

There is a solution to this dilemma[2]. It is to use an intermediary, encoded representation of schedules that is amenable to crossover operations, while employing a decoder that always yields legal solutions to the problem. Let us consider the scheme of representations and decoders that generated the second and third schedules above.

A complete schedule for the job shop was derived from a list of preferences for each work station, linked to times. A preference list had an initial member–a time at which the list went into effect. The rest of the list was made up of some permutation of the contracts available, plus the elements "wait" and "idle". The decoding routine for these representations was a simulation of the job shop's operations, assuming that at any choice point in the simulation, a work station would perform the first allowable operation from its preference list. Thus, if work station 5 had a preference list of the form (60 contract1 contract2 wait idle), and it was minute 60 in the simulation, the simulator looked to see whether there was an object from contract 1 for the work station to work on. If so, that was the task the work station was given to perform. If not, the simulator looked to see whether there was an object from contract 2 to work on. If so, it set the work station to change status to work on contract 2, noting the elapsed time if contract 1 had been worked on last, and then set it to work on the new object. If not, the station waited until an object became available. By moving the "wait" element before contract2, one could cause the work station to process objects from contract 1 only, never changing over to contract 2.

Representing the problem in this way guarantees that legal schedules will be produced, for at each decision point the simulator performs the first legal action contained on a work station's list of all available actions. The decoding routine is a projected simulation, and the evaluation of a schedule is the cost of the work stations, performing the tasks derived in the simulation. As we shall see, the simulation decoder also provides some information that will guide operations to perform useful alterations of a schedule.

6. DETAILS OF OPERATION

The program used a population sized 30, and ran for 20 generations. The problem was tried 20 times. It converged on variations of Schedule 2 14 times and a variation of Schedule 3 6 times[3]. The operations used were derived from those optimizations made by us as we tried to solve the problem deterministically:

RUN-IDLE: If a work station has been waiting for more than an hour, insert a preference list with IDLE as the second member at the beginning of the day, and move the previous initial list to time 60. The probability of applying this operation was the percentage of time the work station spent waiting, divided by the total time of the simulation.

SCRAMBLE: Scramble the members of a preference list. Probability was 5 per cent for each list at the beginning of the run, tapered to 1 per cent at the last generation.

CROSSOVER: Exchange preference lists for selected work stations. Probability was 40 per cent at the beginning of the run, tapered to 5 per cent at the last generation.

Each member of the initial population associated a list of five preference lists with each work station. The preference lists were spaced at one-hour intervals, and each was a random permutation of the legal actions.

The evaluation function summed the costs of simulating the run of the system for five hours with the schedule encoded by an individual. (Although SJS overhead costs are not included in the discussion of the three schedules earlier, the evaluation function included them.) If, at the end of five hours, the contracts were not filled, 1000 was added to the run costs.

7. CONCLUDING OBSERVATIONS

The example discussed above is much simpler than those one would encounter in real life, and the range of operations employed here would have to be widely expanded if a realistic example were approached. In addition, the system here would have to be extended to handle the sorts of phenomena that the ISIS team has handled: establishing connections between levels of abstraction, finding useful operations, and building special constraints into the system, for example.

My belief is that these things could be done if they are successfully done by a deterministic program, for it has been our experience that a quick, powerful way to produce an genetic system for a large search problem is to examine the workings of a good deterministic program in that domain. Wherever the deterministic program produces an optimization of its solution, we include a corresponding operation. Wherever it makes a choice based on some measurement, we make a random choice, using each option's measurement to weight its chances of being selected. The result is a process of mimicry that, if adroitly carried out, produces a system that will out-perform the deterministic predecessor in the same environmental niche.

In the case of the schedules produced above, the genetic operators were just those optimizations of schedules that seemed most beneficial when we attempted to produce good schedules by hand. The crudeness of the approach stems from our lack of any fully specified deterministic solution to more realistic scheduling problems. When fuller descriptions of knowledge-based scheduling routines are available, it will be interesting to investigate their potential for conversion into genetic scheduling systems.

FOOTNOTES

1. "ISIS: A Constraint-Directed Reasoning Approach to Job Shop Scheduling," Mark S. Fox, Bradley P. Allen, Stephen F. Smith, Gary A. Strohm, Carnegie-Mellon University Research Report, 1983.

2. The strategy of encoding solutions in an epistatic domain for operation purposes, while decoding them for evaluation purposes, was worked out and applied to a number of test cases by a group of researchers at Texas Instruments, Inc. The group included me, Nichael Cramer, Garr Lystad, Derek Smith, and Vibhu Kalyan.

3. A number of variations in the scheduling that made no difference in the final evaluation have been omitted in this summary.

Compaction of Symbolic Layout using Genetic Algorithms

Michael P. Fourman
Dept of Electrical and Electronic Engineering
Brunel University, Uxbridge Middx., U.K.
michael%brueer@ucl-cs.AC.UK

Introduction.

Design may be viewed abstractly as a problem of optimisation in the presence of constraints. Such problems become interesting once the space of putative solutions is too large to permit exhaustive search for an optimum, and the payoff function too complex to permit algorithmic solutions. Evolutionary algorithms [Holland 1975] provide a means of guiding the search for good solutions. These algorithms may be viewed as embodying an informal heuristic for problem solving along the lines of

> "To find a better strategy try
> variations on what has worked
> well in the past."

Here, a "strategy" is an attempt at a solution. A strategy will generally not address all the constraints imposed by the problem. The algorithms we are considering guide the search by comparing strategies. We represent this comparison by the relation

a beats b

(which will usually be be a partial order, but need not be total). We call strategies which satisfy all the constraints of the problem "solutions". In general, solutions should beat other strategies and, of course, some solutions will beat others. Abstractly, the algorithms merely search for strategies which are well-placed in this ordering.

Many problems in silicon design involve intractable optimisation problems, for example, partitioning, placement, PLA folding and layout compaction. We say a problem is intractable when the combinatorial complexity of the solution space for the problem makes exhaustive search impossible, and the varied nature of the constraints which must be satisfied makes it unlikely that there is a constructive algorithmic solution to the problem. Automatic solution of such problems requires efficient search of the solution space. Simulated annealing has been applied to the first three problems [Kirkpatrick *et al.* 1983], branch and bound techniques have been applied to layout compaction [Schlag *et al.* 1983]. In this paper we report on the application of a genetic algorithm to layout compaction.

The first prototype solved a highly simplified version of the problem. It produced layouts of a given family of rectangles under the constraint that no two shall overlap, with cost given by the area of a bounding box. A more realistic prototype deals with the layout of a family of rectangular modules with a single level of interconnect. These prototypes allow the designer to add his ideas to the evolving population of layouts and thus supplement rather than replace his expertise.

Symbolic Layout.

A circuit diagram conveys connectivity information:

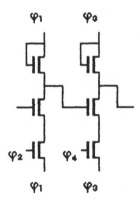

To manufacture the circuit this must be tranformed to a representation in terms of layout elements, each layout element must be assigned an absolute mask position. A layout diagram conveys this mask-making information. The passage from a circuit diagram to a layout may be divided into three stages: firstly the topology (relative positioning of layout elements) of the layout is designed and represented by a **symbolic layout**, then a mask level is assigned to each wire in the circuit – the design is now represented by a **stick diagram**, finally the mask geometry (absolute size and positions) is created. Engineers commonly use these

142

intermediate notations to represent the intermediate stages in the design process.
Here is a mask layout for our circuit:

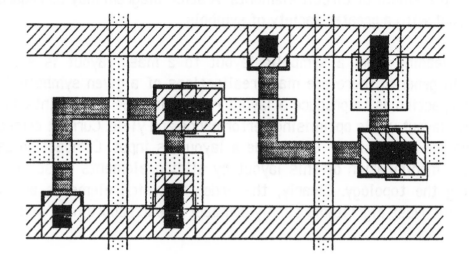

Here is a symbolic version of this layout:

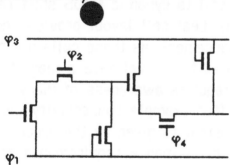

The corresponding stick diagram is:

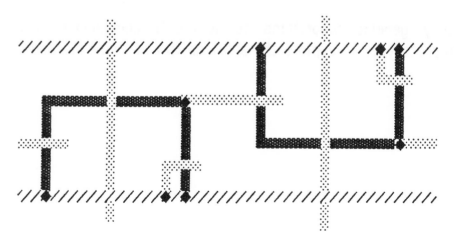

A symbolic layout is a representation of a circuit design which includes some layout information. The symbolic layout represents a number of design decisions on the relative placement of circuit elements. A stick diagram may be regarded as a symbolic layout with a greater variety of symbols.

The procedure leading from a symbolic layout to a mask layout is a form of **compaction**. In general, there are many realisations of a given symbolic layout. The aim of compaction is to produce a layout respecting the constraints implicit in the symbolic layout while optimising performance and yield. Current compaction algorithms require the designer to provide a layout as input. Compaction usually consists of the modification of this layout by sliding elements closer together while retaining the topology. Clearly, the order in which elements are moved affects the result. Most algorithms simply compact in each coordinate direction in turn.

Modern designs are modularised hierarchically. The process of symbolic layout and compaction may occur at any level of this hierarchy. The example we have used for illustration above is a leaf cell (a dynamic NMOS shift register cell) from the bottom level of the hierarchy. Leaf cell layout provides great opportunities for area reduction and yield enhancement, as these cells are replicated many times and any small improvements at this level have a magnified effect on the chip. Optimising leaf cell layout requires awareness of many interacting constraints and complex cost functions (for example connectivity constraints given by the circuit design, geometric constraints given by the process design rules, and the cost functions arising from performance requirements and knowledge of yield hazards). Because of this, constructive algorithmic solutions to this problem have not proved efficient. Traditionally, this area of design has been left to human experts.

We hope to apply genetic algorithms to leaf-cell compaction, and have implemented two prototypes to explore the applicability of these methods in this domain.

144

Genetic Algorithms.

Genetic algorithms are applicable to problems whose solution may be arrived at by a process of successive approximations. This means that we need to be able to modify strategies in such a way that modifications of good strategies are likely to be better than randomly chosen strategies. A simple heuristic in this setting would be to take a strategy, **a**, and randomly generate a modification, M(**a**), of it which may, or may not, be accepted on a probabilistic basis. An algorithm embodying this idea is simulated annealing [Kirkpatrick *et al.* 1983]. The algorithm procedes by starting with a strategy and repeatedly modifying it in this way, varying the acceptance procedure according to the value of a variable called temperature. If M(**a**) beats **a**, the modification is accepted. If **a** beats M(**a**), the modification may be accepted (the probability of this increases with temperature and decreases if M(**a**) is badly beaten). The algorithm is run, starting at a high temperature which is gently lowered. This simulates the mechanism whereby a physical system, gently cooled, tends to reach a low-energy equilibrium position. Genetic algorithms apply where the strategies have more structure. (In fact, in most applications of simulated annealing, this extra structure is available.) Strategies are represented as conjunctions of elementary restrictions on the search space, or **decisions**. The evolutionary algorithm produces a population of strategies, rather than a single strategy. The idea is that by combining some parts of one good strategy with some parts of another, we are likely to produce a good strategy. Thus in generating the progeny of a population, we allow not only modifications or **mutation**, but also **reproduction** which combines part of one strategy with part of another. The basic step is to take a population and produce a number of progeny using a combination of mutation and reproduction. The progeny compete with the older generation, and each other, for the right to reproduce.

If reproduction is to maintain good performance, we need to be able to divide strategies in such a way that decisions which cooperate are likely to stay together. This is accomplished in an indirect and subtle manner. Strategies are represented as strings of decisions. The child, R(**a**,**b**), of **a** and **b** is generated by randomly splitting **a** and **b** and joining part of one to part of the other. Thus, decisions which are close together in the string are likely to stay together. To allow cooperating decisions to become close together, we include **inversions** (which merely choose some substring and reverse it) among the possible mutations. These act together with reproduction and selection, to move decisions which cooperate closer to each other.

Nothing analogous to the temperature used in simulated annealing appears explicitly in the genetic algorithm. The likelihood that a nascent individual will survive to reproduce depends on the degree of competition it experiences from the rest of the population. As the population adapts, the competition hots up –which has the same effect as the cooling in the simulation of annealing.

Although genetic algorithms may be seen as a generalisation of simulated annealing, mutation plays a subsiduary rôle to reproduction. The population at any generation should be viewed as a repository of information summarizing the results of previous evaluations. Individuals which perform well survive to reproduce. Reproduction acts to propagate combinations of decisions occuring in these individuals. The better an individual performs, the longer it will survive and the more chances it has to reproduce. The relative frequencies with which various groups of decisions occur in the population record the degree to which they have been found to work well together. Holland has shown that (under appropriate statistical assumptions) the effect of the genetic algorithm is to use this information to effect an optimal allocation of trials to the various combinations of genes.

Applying the genetic algorithm to compaction.

The genetic algorithm evolves populations of individuals. In our implementation, each individual is characterised by a chromosome which is a string of genes. The length of chromosomes is not fixed. New individuals are produced by a stochastic mix of the classic genetic operators: crossover, mutation and inversion. Crossover picks two individuals at random from the population, randomly cuts their chromosomes and splices part of one with part of the other to form a new chromosome. Mutation picks an individual from the population and, at a randomly chosen number of points in its chromosome, may delete, create or replace a gene. Inversion reverses some substring of a randomly selected chromosome.

A Simple Layout Problem.

The layout problem addressed by our first prototype may be thought of as a form of 2-dimensional binpacking: A collection of rectangles is to be placed in the plane to satisfy certain design rules and minimise some cost function.

The simplest version of this problem (the one we address) has rectangles of fixed sizes, the design rule that distinct rectangles should not overlap, and cost given by area of a bounding box. This version of the problem is already intractable: Suppose we satisfy the constraint that the distinct rectangles, **p,q**, should not overlap, by stipulating that one of the four elementary constraints

p above q
p below q
p left_of q
p right_of q

is satisfied. Then, for a problem with **n** rectangles, we have $N = n^2 - n$ pairs and, *a priori* , 4^N elements in our search space. In fact, this estimate of the size of the problem is unreasonably large, there are ways of reducing the search space significantly; for example, "branch and bound" procedures have been used [Schlag *et al.* 1983].

Layout Strategies.

We consider layout strategies which consist of **consistent** lists of elementary constraints (as above). Given such a list, the rectangles are placed in the first quadrant of the plane as close to the origin as is consistent with the list of elementary constraints. (The procedure which interprets the constraints is very unintelligent. For example, it interprets 'p **above** q' by ensuring that the y-coordinate of the bottom of **p** is greater than that of the top of **q**, even if **p** is actually placed far to the right of **q** (because of other constraints). Any inconsistent lists of constraints produced by the genetic operators are discarded.

Selection criteria.

Populations of consistent lists of constraints are evolved using various orderings for selection. When defining a selection criterion, various conflicting factors must be addressed. For example, our simplest criterion attempts firstly to remove design-rule violations and then to reduce the area of the layout. Strategies with fewer violations beat those with more and, for those with the same number of violations, strategies with smaller bounding boxes win. This simple prioritising of concerns led to the generation of some rather unpromising strategies; while the selection criterion was busy removing design rule violations, for example, any strategy with few such violations (compared to the current population norm) was accepted. Typically, these would have large areas and redundant constraints. The algorithm would later have to spend time refining these crude attempts. In an attempt to mitigate this effect, we added a further selection, favouring shorter chromosomes all other things being equal. Smith has pointed out that implementations of the genetic algorithm allowing variable length chromosomes tend to produce ever longer chromosomes (as chromosomes below a certain length are selected against). We did not find this an overwhelming problem as longer chromosomes were more likely to be rejected as inconsistent by the evaluation function. Nevertheless, we did find that the performance of the algorithm was improved by introducing a selection favouring shorter chromosomes.

We also experimented with trade-offs between the various criteria, established by computing a composite score for each strategy and letting the strategy with the better score win. We found that the genetic algorithm was remarkably robust in optimising the various scoring functions we tried. However, the results were often unexpected; the algorithm would find ways of exploiting the trade-offs provided in unanticipated ways. We have not yet found a selection criterion of this type which works uniformly well, over a range of examples. However, by tuning the selection criterion to the example, good solutions have been obtained.

A better way of combining our various concerns was found. Rather than address the concerns serially, or try to address all the concerns at once, we select a concern randomly each time we have a selection to make. A number of predicates for comparing two individuals were programmed. (For example, comparing areas of bounding boxes, comparing areas of design rule violations, comparing the areas of rectangles placed.) Each time we are asked to compare two individuals, we non-deterministically choose one of these criteria and apply it, ignoring the others. This works surprisingly well. It is easy to code in new criteria and to adjust the algorithm by changing the relative frequencies with which the criteria are chosen. The resulting populations show a greater variability than with deterministic selection, and alleles which perform well in some respects, but would have been selected out with our earlier deterministic approach, are retained.

Results.
Most of our experiments with this prototype have been based on problems with a large amount of symmetry, for which it is easy (for us) to enumerate the optimal solutions. If we actually wanted to solve these problems, other approaches

148

exploiting the symmetries available would certainly be more efficient. However, for the purpose of evaluating the performance of the genetic algorithm, we claim these examples are not too misleading. The algorithm is not provided with any knowledge of the symmetries of the problem nor of the arithmetical relationships between the sizes of the rectangles. For the purposes of evaluating the applicability of the genetic algorithm to layout compaction, the prototype is probably pessimistic. Real layout problems are far more constrained (by, for example, connectivity constraints). This not only reduces the size of the search space *per se*, but also appears to localise the interdependence of various genes making the problem more suitable for the genetic algorithm.

A naive analysis of a very simple example is instructive. The example consists of six rectangles, three 3 x 1 (horizontal) and three 1 x 3 (vertical). A minimal solution of this problem was found (consistently) in under 50 generations with 20 progeny per generation (1000 points of the search space evaluated).

A solution to this problem must say how each of these rectangles is constrained, both horizontally and vertically. Thus the search space has 6^{12} (about 2×10^9) points. The problem has 8 basic solutions and a symmetry group of order 36. There are about 7.5×10^6 points/solution. Of these, we only examine some 10^3.

Representing Layout.

Our first prototype deals with a problem which has little direct practical significance for VLSI layout. (However, Rob Holte has pointed out that scheduling problems from operations research might be represented by minor variations on our prototype problem.) As a next step towards a practical layout tool, we have implemented a system which compacts a simple form of symbolic layout. The problem is to formalise the constraints implicit in the symbolic layout, and to find a representation, suitable for the genetic algorithm, for layout strategies.

We consider a symbolic layout of blocks connected by wires. The rectangles (blocks) are of fixed size and may be translated but not rotated. The interconnecting lines (wires) are of fixed width but variable length. The interconnections shown must be maintained, and no others are allowed. In addition, there are design rules which prohibit unconnected pairs of tiles (wires or blocks) from being placed too close together.

This form of the symbolic layout problem was introduced by [Schlag *et al.* 1983]. Here is their example of a simple symbolic layout:

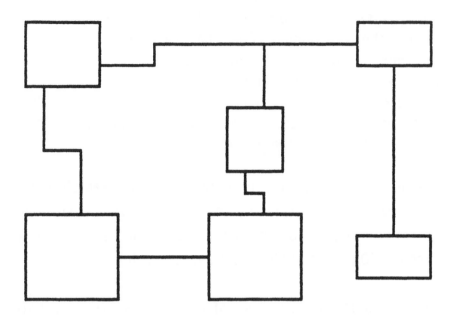

We represent the problem at two levels:

A surface level deals with **tiles** of three kinds - blocks, horizontal wires and vertical wires. In addition to evolving layout constraints dealing with the relative positions of tiles (above, right-of etc. as before), we use a fixed list of **structural constraints**, to represent the information in the symbolic layout, and **fundamental constraints** which represent the size limitations on tiles. Structural constraints have the following forms

v crosses h, N b v, S b v, E b h, W b h

where **v, h** are vertical and horizontal wires and **b** is a block. These constraints allow us to stipulate which wires cross (and hence are connected) and which wires connect to which edges (North, South, East or West) of which blocks.

At a deeper level, unseen by the user, the problem is represented in terms of the **primitive layout elements**, north b, south b, east b, west b, left h, right h, y_posn h, top v, btm v, x_posn v, whose names are self-explanatory. For each tile, we generate a list of fundamental constraints expressing the relationship between the primitive layout elements arising from it. This representation allows both blocks and wires to stretch.

The example above is represented by declaring the widths of the wires and sizes of the blocks and then specifying the following list of constraints. (We use a LISP list syntax as it is more widely familiar, actually, our implementation is written in ML.):

```
((E B1 H2)
(crosses V3 H2)
(crosses V3 H3)
(crosses V4 H3)
(N B4 V4)
(W B5 H3)

(S B1 V1)
(crosses V1 H1)
(crosses V2 H1)
(N B2 V2)

(E B2 H5)
(W B3 H5)

(S B4 V6)
(crosses V6 H4)
(crosses V5 H4)
(N B3 V5)

(S B5 V7)
(N B6 V7))
```

Again, we evolve lists of layout constraints. These are compiled, together with the fixed structural and fundamental constraints representing the symbolic layout to give graphs of constraints on the primitive layout elements, whose positions are thus determined. The number of design-rule violations and the area of the resulting layout are again used to select between rival strategies. Solutions to this problem were found in around 200 generations of 20 progeny, and this was reduced to around 150 generations when the algorithm was given a few "hints" in the form of extra constraints. Watching the evolving populations showed that progress was rapid for around 50 generations. Thereafter, the algorithm appeared to get stuck for long periods on local minima (in the sense that one configuration would dominate the population). This lack of variation in the population reduced the usefulness of crossover. When mutation led to a promising new configuration, there would be a period of experimentation leading rapidly to a new local minimum. This might suggest that either the population size (100) or the probability of mutation being used as an operator (0.1) is too small. We have not yet experimented with variations on these parameters. We think that better solutions would be either to introduce a further element of competition into the genetic algorithm by penalising configurations which become too numerous (implementing this is problemaical), or to evolve a number of populations allowing a limited degree of "intermarriage" (We are currently implementing the latter approach. If it is successful it will be a good candidate for parallel implementation.)

Conclusions.

The genetic algorithm may be viewed as a (non-deterministic) machine which is programmed by supplying it with a selection criterion - an algorithm for comparing two lists of constraints. We have experimented with various selection criteria based on combinations of the total intersection area, I, of overlap involved in design-rule violations, and the area, A, of a bounding rectangle. Experiments were made to compare various performance criteria based on combinations of the number of design-rule violations, and the area of a bounding rectangle. From our experience with the prototype, it appears that the choice of a selection criterion is an essential difficulty in applying the genetic algorithm to layout. The problem is that we must evolve populations of partial solutions (strategies), while the optimisation task is defined in terms of a cost function defined on layouts (solutions). To extend a (technology imposed) cost-function, defined on solutions, to the space of strategies, in such a way that the genetic algorithm will produce a solution (rather than just a high-scoring strategy), is a non-trivial task.

We intend to experiment with our second prototype in various ways before going on to implement a "real" system dealing with design-rules for a practical multi-layer technology. We will continue to experiment with selection criteria and we are implementing the idea of having several weakly interacting populations running in parallel, described above. We also intend to integrate other, rule-based, methods with the genetic algorithm, automating the provision of "hints". Thus, a number of suggestions for strategies would be generated and passed to the genetic algorithm which would then explore combinations and variations of these.

Acknowledgements.

I would like to thank Steve Smith for introducing me to Genetic Algorithms, and Robert Holte for many stimulating discussions, his criticism and encouragement have been invaluable.

References.

Holland, John H. 1975. Adaptation in natural and artificial systems. Ann Arbor, University of Michigan Press.

Kirkpatrick, S., C.D. Gelatt, and M.P. Vecchi 1983. Optimisation by simulated annealing. *Science,* 1983, 220, 671-680.

Schlag, M., Y.-Z. Liao, and C.K. Wong 1983. An algorithm for optimal two-dimensional compaction of VLSI layouts. *INTEGRATION, the VLSI journal* 1 (1983) 179-209.

Smith, S.F. 1982. Implementing an adaptive learning system using a genetic algorithm. Ph.D. thesis. U. of Pittsburgh 1982.

ALLELES, LOCI, AND THE TRAVELING SALESMAN PROBLEM

by

David E. Goldberg

and

Robert Lingle, Jr.

Department of Engineering Mechanics
The University of Alabama, University, AL 35486

INTRODUCTION

We start this paper by making several seemingly not-too-related observations:

1) Simple genetic algorithms work well in problems which can be coded so the underlying building blocks (highly fit, short defining length schemata) lead to improved performance.

2) There are problems (more properly, codings for problems) that are GA-Hard --difficult for the normal reproduction+crossover+mutation processes of the simple genetic algorithm.

3) Inversion is the conventional answer when genetic algorithmists are asked how they intend to find good string orderings, but inversion has never done much in empirical studies to date.

4) Despite numerous rumored attempts, the traveling salesman problem has not succumbed to genetic algorithm-like solution.

Our goal in this paper is to show that, in fact, these observations are closely related. Specifically, we show how our attempts to solve the traveling salesman problem (TSP) with genetic algorithms have led to a new type of crossover operator, partially-mapped crossover (PMX), which permits genetic algorithms to search for better string orderings while still searching for better allele combinations. The partially-mapped crossover operator combines a mapping operation usually associated with inversion and subsequent crossover between non-homologous strings with a swapping operation that preserves a full gene complement. The resultant is an operator which enables both allele and ordering combinations to be searched with the implicit parallelism usually reserved for allele combinations in more conventional genetic algorithms.

In the remainder, we first examine and question the conventional notions of gene and locus. This leads us to consider the mechanics of the partially-mapped crossover operator (PMX). This discussion is augmented by the presentation of a sample implementation (for ordering-only problems) in Pascal. Next, we consider the effect of PMX by extending the normal notion of a schema by introducing the o-schemata (ordering schemata) or locus templates. This leads to simple counting arguments and survival probability calculations for o-schemata under PMX. These results show that with high probability, low order o-schemata survive PMX thus giving us a desirable result: an operator which searches among both orderings and allele combinations that lead to good fitness. Finally, we demonstrate the effectiveness of this extended genetic algorithm consisting of reproduction+PMX, by applying it to an ordering-only problem, the traveling salesman problem (TSP). Coding the problem as an n-permutation with no allele values, we obtain optimal or very near-optimal results in a well-known 10 city problem. Our discussion concludes by discussing extensions in problems with both ordering and value considered.

THE CONVENTIONAL VIEW OF POSITION AND VALUE

In genetic algorithm work we usually take a decidedly Mendelian view of our artificial chromosomes and consider genes which may take on different values (alleles) and positions (loci). Normally we assume that alleles decode to our problem parameter set (phenotype) in a manner independent of locus. Furthermore, we assume that our parameter set may then be evaluated by a fitness function (a non-negative objective function to be maximized). Symbolically, the fitness f depends upon the parameter set x which in turn depends upon the allele values v or more compactly $f = f(x(v))$. While this is certainly conventional, we need to ask whether this is the most general (or even most biological) way to consider this mapping. More to the point, shouldn't we also consider the possible effect of a string's ordering o on phenotype outcome and fitness. Mathematically there seems to be no good reason to exclude this possibility which we may write as $f=f(x(o,v))$.

While this generalization of our coding techniques is attractive because it would permit us to code ordering problems more naturally, we must make sure we maintain the implicit parallelism of the reproductive plans and genetic operators we apply to the generalized structures. Furthermore, because GA's are drawn from biological example we should be careful to seek natural precedent before committing ourselves to this extension. To find biological precedent for the importance of ordering as well as value we need only consider the sublayer of structure beneath the chromosome and consider the amino acid sequences that lead to particular proteins. At this level, the values (amino acids) are in no way tagged with meaning. There are only amino acids and they must appear in just the right order to obtain a useful outcome (a particular protein). Thus, there is biological example of outcomes that depend upon both ordering and value, and we do not risk the loss of the right flavor by considering them both.

Then, wherein lies our problem? If it is ok to admit both ordering and value information into our fitness evaluation, what is missing in our current thinking about genetic algorithms which prevents us from exploiting both ordering and value information concurrently? In previous work where ordering was considered at all (primarily for its effect on the creation of good, tightly linked, building blocks), the only ordering operator considered was inversion, a unary operator which picks two points along a single string at random and inverts the included substring (1). Subsequent crossover between non-homologous (differently ordered) strings occurred by mapping one string's order to the other, crossing via simple crossover, and unmapping the offspring. This procedure is well and good for searching among different allele combinations, but it does little to search for better orderings. Clearly the only operator effecting string order here is inversion, but the beauty of genetic algorithms is contained in the structured, yet randomized information exchange of crossover--the combination of highly fit notions from different strings. With only a unary operator to search for better string orderings, we have little hope of finding the best ordering, or even very good orderings, in strings of any substantial length. Just as mutation cannot be expected to find very good allele schemata in reasonable time, inversion cannot be expected to find good orderings in substantial problems. What is needed is a binary, crossover-like operator which exchanges both ordering and value information among different strings. In the next section, we present a new operator which does precisely this. Specifically, we outline an operator we call partially-mapped crossover (PMX) that exploits important similarities in value and ordering simultaneously when used with an appropriate reproductive plan.

PARTIALLY-MAPPED CROSSOVER (PMX) - MECHANICS

To exchange ordering and value information among different strings we present a new genetic operator with the proper flavor. We call this operator partially-mapped crossover because a portion of one string ordering is mapped to a portion of another and the remaining information is exchanged

after appropriate swapping operations. To tie down these ideas we also present a piece of code used in the computational experiments to be presented later.

To motivate the partially-mapped crossover operator (PMX) we will consider different orderings only and neglect any value information carried with the ordering (this is not a limitation of the method because allele information can easily be tacked on to city name information). For example, consider two permutations of 10 objects:

```
A = 9 8 4 5 6 7 1 3 2 10
B = 8 7 1 2 3 10 9 5 4 6
```

PMX proceeds as follows. First, two positions are chosen along the string uniformly at random. The substrings defined from the first number chosen to the second number chosen are called the MAPPING SECTIONS. Next, we consider each mapping section separately by mapping the other string to the mapping section through a sequence of swapping operations. For example, if we pick two random numbers say 4 and 6, this defines the two mapping sections, 5-6-7 in string A, and 2-3-10 in string B. The mapping operation, say B to A, is performed by swapping first the 5 and the 2, the 6 and the 3, and the 7 and the 10, resulting in a well defined offspring. Similarly the mapping and swapping operation of A to B results in the swap of the 2 and the 5, the 3 and the 6, and the 10 and the 7. The resulting two new strings are as follows:

```
A' = 9 8 4 2 3 10 1 6 5 7
B' = 8 10 1 5 6 7 9 2 4 6
```

The mechanics of PMX is a bit more complex than simple crossover so to tie down the ideas completely we present a code excerpt which implements the operator for ordering-only structures in Figure 1. In this code, the string is treated as a ring and attention is paid to the order of selection of the two mapping section endpoints.

The power of effect of this operator, as with simple crossover, is much more subtle than is suggested by the simplicity of the string matching and swapping. Clearly, however, portions of the string ordering are being propagated untouched as we should expect. In the next section, we identify the type of information being exchanged by introducing the o-schemata (ordering schemata). We also consider the probability of survival of particular o-schemata under PMX.

PARTIALLY-MAPPED CROSSOVER - POWER OF EFFECT

In the analysis of a simple genetic algorithm with reproduction+crossover+mutation, we consider allele schemata as the underlying building blocks of future solutions. We also consider the effect of the genetic operators on the survivability of

Data Types and Constants

```
const max_city = 100;

type  city     = 1..max_city;
      tourarray = array[1..max_city] of city;
```

Functions and Procedures (find_city, swap_city, cross_tour)

```
function find_city(city_name,n_city:city; var tour:tourarray):city;
var j1:integer;
begin
 j1:=0;
 repeat
  j1:=j1+1;
 until ( (j1>n_city) or (tour[j1]=city_name) );
 find_city:=j1;
end;

procedure swap_city(city_pos1,city_pos2:integer; var tour:tourarray);
var temp:city;
begin
 temp:=tour[city_pos1];
 tour[city_pos1]:=tour[city_pos2];
 tour[city_pos2]:=temp;
end;

procedure cross_tour(n_city,lo_cross,hi_cross:city;
                     var tour1_old,tour2_old,tour1_new,tour2_new:tourarray);
var j1,hi_test:integer;
begin       .
 hi_test := hi_cross + 1; if (hi_test>n_city) then hi_test:=1;
 tour1_new := tour1_old;
 tour2_new := tour2_old;
 if ( (lo_cross <> hi_cross) and (lo_cross <> hi_test) ) then begin
   j1 := lo_cross;
   while (j1<>hi_test) do begin (* mapped crossover on both tours *)
    swap_city(j1,find_city(tour1_old[j1],n_city,tour2_new),tour2_new);
    swap_city(j1,find_city(tour2_old[j1],n_city,tour1_new),tour1_new);
    j1:=j1+1; if (j1>n_city) then j1:= 1;
   end;
  end;
end;
```

Figure 1. Pascal Implementation of PMX - Partially Mapped
Crossover - procedure cross_tour.

important schemata. In a similar way, in our current work we consider the o-schemata or ordering schemata, and calculate the survival probabilities of important o-schemata under the PMX operator just discussed. As in the previous section we will neglect any allele information which may be carried along to focus solely on the ordering information; however, we recognize that we can always tack on the allele information for problems where it is needed in the coding.

To motivate an o-schema consider two of the 10-permutations:

```
C = 1 2 3 4 5 6 7 8 9 10
D = 1 2 3 5 4 6 7 9 8 10
```

As with allele schemata (a-schemata) where we appended a * (a meta-don't-care symbol) to our k-nary alphabet to motivate a notation for the schemata or similarity templates, so do we here append a don't care symbol (the !) to mean that any of the remaining permutations will do in the don't care slots. Thus, in our example we have, among others, the following o-schemata common among structures C and D:

```
1 2 3 ! ! ! ! ! ! !
! 2 ! ! ! ! ! ! ! !
! ! ! ! ! 6 7 ! ! !
1 ! ! ! ! 6 ! ! ! 10
```

To consider the number of o-schemata, we take them with no positions fixed, 1 position fixed, 2 positions fixed, etc., and recognize that the number of o-schemata with exactly j positions fixed is simply the product of the number of combinations of groups of j among ℓ objects, $\binom{\ell}{j}$, times the number of permutations of groups of j among ℓ objects. Summing from 0 to ℓ (the string length) we

156

obtain the number of o-schemata:

$$n_{os} = \sum_{0}^{\ell} \frac{\ell!}{(\ell-j)!j!} \frac{\ell!}{(\ell-j!)}$$

While this expression has not been reduced to closed form, it may be shown for large ℓ that the number of o-schemata is certainly greater than $(\ell!)^2$. Furthermore, it is easily shown that each particular string (permutation) is a representative of 2^{ℓ} o-schemata and that a population contains at most $n \cdot 2^{\ell}$ o-schemata.

Next we consider the survival probability of a particular o-schema under the partially-mapped crossover operator. The easiest way to calculate this is to use conditional probabilities over three mutually exclusive events: the o-schema is entirely contained within the match section (Event W-within), the schema is entirely outside the match section (Event O-outside), or the schema is cut by a cross point (Event C-cut). Thus, the probability of survival (Event S-survival) may be given:

$$P(S) = P(S|W)P(W) + P(S|O)P(O) + P(S|C)P(C)$$

Since the probability of surviving a cut is very low $(P(S|C) \cong 0)$ we ignore this possibility and focus on the other two events. Assuming a cut length k, a defining length of the schema $\delta(s)$, and an o-schema of order (number of fixed positions) $o(s)$, the overall probability of survival (for large string length ℓ) may be estimated:

$$P(S) = \frac{k-\delta+1}{\ell} + \frac{\ell-k-\delta+1}{\ell} \left(1 - \frac{k+1}{\ell}\right)$$

Closer examination of this equation reveals two modes of survival. When the cut length is large with respect to the defining length, relatively short defining length schemata survive with high probability. The second and more subtle mode of survival occurs when short, low order schemata survive, because a small cut length dictates a small probability of interruption due to swapping. Together the two modes combine to pass through short, low order o-schemata so normal reproductive plans can sample these building blocks at near-optimal rates. Hence, PMX permits the same type of implicit parallelism to occur in both orderings and alleles as we have already witnessed using simple crossover on allele information alone.

A PURE ORDERING PROBLEM - THE TRAVELING SALESMAN PROBLEM (TSP)

In some sense we've presented this paper in the reverse order of discovery. We did not 1) admit ordering information, 2) discover PMX and o-schemata, and 3) apply reproduction+PMX to the traveling salesman problem. In fact, by trying to solve the TSP with genetic algorithms, we were led to PMX-like operators, then o-schemata, and

finally PMX. The traveling salesman problem is a pure ordering problem (2,3,4) where one attempts to find the optimal tour (minimum cost path which visits each of n cities exactly once). The TSP is possibly the most celebrated combinatorial optimization problem of the past three decades, and despite numerous exact (impractical) and heuristic (inexact) methods already discovered, the method remains an active research area in its own right, partially because the problem is part of a class of problems considered to be NP-complete for which no polynomial time solution is believed to exist. Our interest in the TSP sprung mainly from a concern over claims of genetic algorithm robustness. If GA's are robust, why have the rumored attempts at "solving" the TSP with GA's failed. This concern led us to consider many schemes for coding the ordering information, with strange codes, penalty functions, and the like, but none of these had the appropriate flavor--the building blocks didn't seem right. This led us to consider the current scheme, which does have appropriate building blocks, and as we shall soon see, does (in one problem) lead to optimal or near-optimal results.

The specific problem we consider is Karg and Thompson's well-studied 10 city problem (4). While a 10 city problem is no final touchstone of success, it does contain 9! alternatives (the GA knows nothing of the problem's symmetry which reduces this number to (9!)/2). We code the problem as a normalized (city 1 in the first position) 10-permutation and apply reproduction and PMX to successive populations. We use roulette wheel reproduction with selection probabilities set in the normal way, and fitnesses are created from costs and scaled by subtracting string cost from population maximum cost, $f_i = c_{max} - c_i$. We choose initial populations, popsize=200, at random. This number was selected to obtain a rich spread of order 2 o-schemata in the population. This requires a population size proportional to $n(n-1)$ or roughly n^2. It might be useful to have order 3 schemata as well, but this may require larger populations than we are used to working with.

We present the results of two runs on the 10 city problem in Figures 2 and 3. Figure 2 shows the population average cost with each successive generation. The crossover probability was set at 0.6 so each generation represents roughly 120 new function evaluations (0.6*200). Figure 3 shows the population best results with successive generations. As we can see, run 1 reaches the optimal (!!) result rather quickly, while run 2 converges on a very near-optimal tour (we only ran twenty generations--there was still enough diversity left so improvement was possible in run 2). The best of run 1 was indeed the Karg and Thompson optimum, tour 1-2-3-4-5-10-9-8-6-7 with cost=378. The best of run 2 was a near-optimum, the tour 1-2-3-10-9-5-4-6-8-7 with cost=381. We are

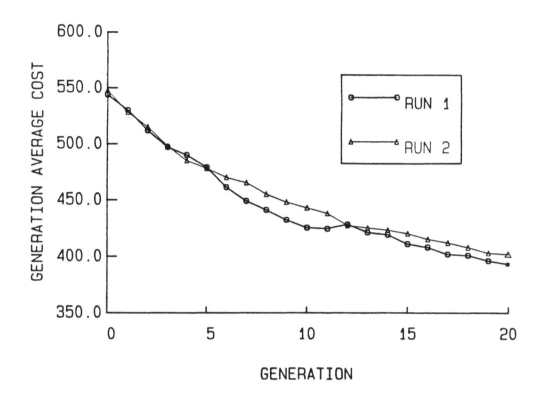

Figure 2. Generation Average Cost vs. Generation for 10 City TSP

currently working on a 20 city problem and a 33 city problem, although we need to do some reprogramming to fit the large population sizes into our IBM PC's. We also have built in an inversion operator, but have not had a chance to test its effect on average and best results.

CONCLUSIONS

In this paper we have examined a new type of crossover operator, partially-mapped crossover (PMX), for the exploration of codings where ordering and allele information may directly or indirectly effect fitness values. The mechanics of the operator have been described, and an ordering-only implementation has been presented in Pascal. The power of effect of the new operator has been analyzed using an extension to the concept of schemata called the o-schemata (ordering schemata). Simple counting arguments have been put forward which show the vast amount of information contained in the o-schemata, and survival probabilities have been estimated for o-schemata under the PMX operator. The result is an operation which preserves ordering building blocks (and allele building blocks if they are attached) so orderings and allele combinations may be explored with implicit parallelism.

The new operator is tested in an ordering-only problem, the traveling salesman problem. Using reproduction+PMX in two runs, optimal or very near optimal results are found in a well-known 10 city problem after exploring a small portion of the tour search space. We are continuing our work by testing the method in larger problems, but we are encouraged with the GA-like performance obtained on our first test.

This work has important implications for improving more general GA-search in problems where both allele combinations and ordering information are important. The binary operation of PMX does permit the randomized, yet structured, information exchange among both alleles and ordering building blocks which simple crossover promotes among allele schemata alone. This should assist us in our efforts to successfully apply genetic algorithms to ever more complex problems.

REFERENCES

1. Holland, J. H., _Adaptation in Natural and Artificial Systems_, University of Michigan Press, Ann Arbor, 1975.

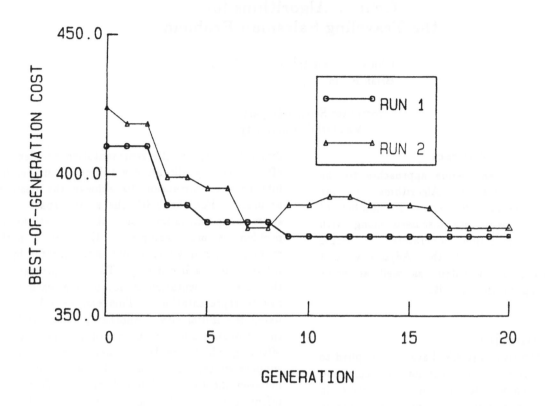

Figure 3. Best-of-Generation Cost for 10 City TSP

2.   Bellmore, M. and G. L. Neuhauser, "The
     Traveling Salesman Problem: A Survey,"
     Operation Research, vol. 16, May-June
     1968, pp. 538-558.

3.   Parker, R. G. and R. L. Rardin, "The
     Traveling Salesman Problem: An Update
     of Research," Naval Research Logistics
     Quarterly, vol. 30, 1983, pp. 69-96.

4.   Karg, R. L. and G. L. Thompson, "A
     Heuristic Approach to Solving Travelling
     Salesman Problems," Management Science,
     vol. 10, no. 2, January 1964, pp.
     225-248.

Genetic Algorithms for
the Traveling Salesman Problem

John Grefenstette[1], Rajeev Gopal,
Brian Rosmaita, Dirk Van Gucht

Computer Science Department
Vanderbilt University

Abstract

This paper presents some approaches to the application of Genetic Algorithms to the Traveling Salesman Problem. A number of representation issues are discussed along with several recombination operators. Some preliminary analysis of the Adjacency List representation is presented, as well as some promising experimental results.

1. Introduction

Genetic Algorithms (GA's) have been applied to a variety of function optimization problems, and have been shown to be highly effective in searching large, complex response surfaces even in the presence of difficulties such as high-dimensionality, multimodality, discontinuity and noise [4]. However, GA's have not been applied extensively to combinatorial problems. The major obstacle is in finding an appropriate representation. This paper presents some approaches to the design of GA's for a well known combinatorial optimization problem – the Traveling Salesman Problem (TSP). The TSP is easily stated: Given a complete graph with N nodes, find the shortest Hamiltonian path through the graph. (In this paper, we will assume Euclidean distances between nodes.) The TSP is NP-Hard, which probably means that any algorithm which computes an exact solution of the TSP requires an amount of computation time which is exponential in N, the size of the problem [5]. In addition to its many important applications, the TSP is often used to illustrate heuristic search methods [2,7,8], so it is natural to investigate the use of GA's for this problem.

Choosing an appropriate representation is the first step in applying GA's to any optimization problem. If the problem involves searching an N-dimensional space, the representation problem is often solved by allocating a sufficient number of bits to each dimension to achieve the desired accuracy. For the TSP, the search space is a space of permutations and the representation problem is more complex. Consider a path representation in which a tour is represented by a list of cities: (a b c d e f). The first problem is that the representation is not unique: each tour has N representations. This can be solved by fixing the initial city. Another problem is that the crossover operator does not generally yield offspring which are legal tours. For example, suppose we cross tours (a b c d e) and (a d e c b) between the third and fourth cities. We get as offspring (a b c c b) and (a d e d e), neither of which are legal tours. Finally, there is a problem in applying the hyperplane analysis of GA's to this representation. The definition of a hyperplane is unclear in this representation. For example, (a # # # #) appears to be a first order hyperplane, but it contains the entire space. The problem is that in this representation, the semantics of an allele in a given position depends on the surrounding alleles. Intuitively, we hope that GA's will tend to construct good solutions by identifying good building blocks and eventually combining these to get larger building blocks. For the TSP, the basic building blocks are edges. Larger building blocks correspond to larger subtours. The path representation does not lend itself to the description of edges and longer subtours in ways which are useful to the GA.

In section 2, we present two representations which offer some improvements over the path representation. Section 3 discusses the design of a heuristic recombination operator for what we consider to be the most promising representation. In section 4, some preliminary experimental

[1] Research supported in part by the National Science Foundation under Grant MCS-8305693.

160

results are described for the TSP. Section 5 discusses some future directions.

2. Representations for TSP

2.1. Ordinal Representation

In the ordinal representation, a tour is described by a list of N integers in which the ith element can range from 1 to (N-i+1). Given a path representation of a tour, we can construct the ordinal representation TourList as follows: Let FreeList be an ordered list of the cities. For each city in the tour, append the position of that city in the FreeList to the TourList and delete that city from the FreeList. For example, the path tour (a c e d b) corresponds to an ordinal tour (1 2 3 2 1) as shown:

TourList	FreeList
()	(a b c d e)
(1)	(b c d e)
(1 2)	(b d e)
(1 2 3)	(b d)
(1 2 3 2)	(b)
(1 2 3 2 1)	()

Note that it is necessary to fix the starting city to avoid multiple representation of tours.

A similar procedure provides a mapping from the ordinal representation back to the path representation. In fact, the mapping between the two representations is one-to-one.

The primary advantage of the ordinal representation is that the classical crossover operator may be freely applied to the ordinal representation and will always produce the ordinal representation of a legal tour. However, the results of crossover may not bear much relation to the parents when translated to the path representation. For example, consider the following two tours:

ordinal tours	path tours
(1 2 3 2 1)	(a c e d b)

(2 4 1 1 1)	(b e a c d)

Suppose that we cross the ordinal tours between the second and third positions. We get the following tours as offspring:

ordinal tours	path tours
(1 2 1 1 1)	(a c b d e)
(2 4 3 2 1)	(b e d c a)

The subtours corresponding to the genes in the ordinal tours to the left of the crossover point do not change. However, the subtours corresponding to genes to the right of the crossover points are disrupted in a fairly random way. Furthermore, the closer the crossover point is to the front of the tour, the greater the disruption of subtours in the offspring.

As predicted by the above consideration of subtour disruptions, experimental results using the ordinal representation have been generally poor. In most cases, a GA using the ordinal representation does no better than random search on the TSP.

2.2. Adjacency Representation

In the adjacency representation, a tour is described by a list of cities. There is an edge in the tour from city i to city j iff the allele in position i is j. For example, the path tour (1 3 5 4 2) corresponds to the adjacency tour (3 1 5 2 4). Note that any tour has exactly one adjacency list representation.

2.2.1. Crossover Operators

Unlike the ordinal representation, the adjacency representation does not allow the classical crossover operator. Several modified crossover operators can be defined.

Alternating Edges

Using the alternating edges operator, an offspring is constructed from two parent tours as follows: First choose an edge at random from one parent. Then extend the partial tour by choosing the appropriate edge from the other parent.

Continue extending the tour by choosing edges from alternating parents. If the parent's edge would introduce a cycle into a partial tour, then extend the partial tour by a random edge which does not introduce a cycle. Continue until a complete tour is constructed.

For example, suppose we have

mom = (2 3 4 5 6 1)
dad = (2 5 1 6 4 3)

Then we might get the following offspring:

kid = (2 5 4 1 6 3)

where the only random edge introduced into the offspring is the edge (4 1). All other edges were inherited by alternately choosing edges from parents, starting with the edge (1 2) from mom.

Experimental results with the alternating edges operator have been uniformly discouraging. The obvious explanation seems to be that good subtours are often disrupted by the crossover operator. Ideally, an operator ought to promote the development of coadapted alleles, or in the TSP, longer and longer high performance subtours. The next operator was motivated by the desire to preserve longer parental subtours.

Subtour Chunks

Using the subtour chunking operator, an offspring is constructed from two parent tours as follows: First choose a subtour of random length from one parent. Then extend the partial tour by choosing a subtour of random length from the other parent. Continue extending the tour by choosing subtours from alternating parents. During the selection of a subtour from a parent, if the parent's edge would introduce a cycle into a partial tour, then extend the partial tour by a random edge which does not introduce a cycle. Continue until a complete tour is constructed.

Subtour chunking performed better than alternating edges, as expected, but the absolute performance was still unimpressive. An analysis of the allocation of trials to hyperplanes provide a partial explanation for the poor performance of this operator.

2.2.2. Hyperplane Analysis

The primary advantage of the adjacency representation is that it permits the kind of hyperplane analysis which has been applied to the N-dimensional function optimization GA paradigm [1,3,6]. Hyperplanes defined in terms of a single defining position correspond to the natural building blocks, i.e., edges, for the TSP problem. For example, the hyperplane (# # # 2 #) is the set of all permutations in which the edge (4 2) occurs. We briefly summarize the main points of the classical hyperplane analysis of GA's: In the absence of recombination operators, selection of structures for reproduction in proportion to the structure's observed relative performance allocates trials to all represented hyperplanes in the population (roughly) according to the following formula:

$$M(H,t+1) = M(H,t)*(u(H,t) / u(P,t))$$

where

$M(H,t) = $ # of representatives of H at time t

$u(H,t) = $ observed performance of H at time t

$u(P,t) = $ mean performance of population at time t.

The elements of any hyperplane partition compete against the other elements of that partition, with the better performing elements eventually propagating through the population. This in turn leads to a reduction in the dimensionality of the search space, and the construction of larger high performance building blocks.

In the adjacency representation, a first order hyperplane partition consists of all of the hyperplanes which are defined on the same position. For example:

{ (# # # 1 #), (# # # 2 #), (# # # 3 #),
(# # # 5 #) }

is a first order hyperplane partition. Each element of the partition contains an equal

162

number of tours. Selection is supposed to distinguish among the elements of this partition and to favor the high performance hyperplanes. However, the following theorem shows that selection has very little information on which to allocate trials to competing first order hyperplanes.

Theorem 1. Suppose that H_{ab} and H_{ac} are two first-order hyperplanes defined by the edges (a b) and (a c), respectively, in a Euclidean TSP. Then $| u(H_{ab}) - u(H_{ac}) | \leq 4(ab + ac)$ where ab and ac represent the lengths of the edges (a b) and (a c), respectively.

Proof. We show that there is a one-to-one mapping f between the tours in H_{ab} and the tours H_{ac} such that if x is a tour in H_{ab} and $y = f(x)$ is the corresponding tour in H_{ac}, then

$$| \text{Length}(y) - \text{Length}(x) | \leq 4(ab+ac).$$

The theorem follows directly.

The following illustrates the mapping f:

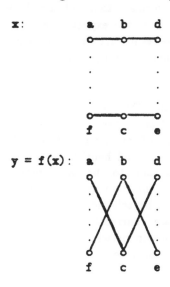

That is, y is obtained by exchanging the nodes b and c in the tour x. Using the triangle inequality, it is easy to show that:

$$-(4ab + 2ac) \leq \text{Distance}(y) - \text{Distance}(x) \leq (4ac + 2ab).$$

So

$$| \text{Distance}(y) - \text{Distance}(x) | \leq 4(ab+ac).$$
QED.

In practice, the observed difference between competing first order hyperplanes is usually an order of magnitude less than the bounds in the theorem. And since the overall tour length is generally very large compared to the bound in the theorem, there is generally no significant difference between the mean relative performance of any two competing first order hyperplanes. Our experimental studies have shown that the difference in the observed performance of competing first order hyperplanes in a TSP of size 20 is generally less than 5% of the mean population tour length. In larger problems, this difference can be expected to rapidly approach zero.

One might suspect that the TSP is not a suitable problem for GA's, that the TSP is in some sense GA-Hard. Bethke[1] characterizes some problems for which GA's are unsuitable. Informally, Bethke shows that there are functions and representations for which the low order hyperplanes can mislead the GA into allocating trials to suboptimal areas of the search space. However, Bethke's techniques, which involve the Walsh transform of the objective function, apply to one-dimensional functions of a real variable using a fixed-point representation. A similar set of results may be derivable for combinatorial problems using the adjacency representation. But Theorem 1 does not indicate that the information in the first order hyperplanes of the adjacency representation is misleading, just that it is buried. In other words, measuring the fitness of a tour by the tour length may be too crude a measure for apportioning credit. We now describe a crossover operator which performs a secondary apportionment of credit at the level of individual alleles.

3. Heuristic Crossover

Theorem 1 shows that selection alone may not be able to properly allocate trials to first order hyperplanes, given our adjacency representation for the TSP. The heuristic crossover operator attempts to perform a secondary apportionment of credit at the allele level. This operator constructs an offspring from two parent tours as follows: Pick a random city as the starting point for the child's tour. Compare the two edges leaving the starting city in the parents and choose the shorter edge. Continue to extend the partial tour by choosing the shorter of the two edges in the parents which extend the tour. If the shorter parental edge would introduce a cycle into the partial tour, then extend the tour by a random edge. Continue until a complete tour is generated.

In order to compare this operator with the previous two recombination operators, 1000 random pairs of parents were chosen for a TSP of size 20. For each pair of parents, an offspring was constructed according to each of the crossover operators. For all three operators, the offspring generally inherited about 30% of the edges from each parent. The remaining 40% were random edges introduced by the recombination operator to create a legal tour. For the first two operators, the offspring generally show no improvement in overall tour length when compared to the better parent. Not surprisingly, the heuristic crossover produces offspring which are, on average, about 10% better than the better parent. It seems reasonable that such an improvement should give selection a way to promote the propagation of good edges through the population. The next section shows some experimental results which confirm this expectation.

It is important to note that, with the proper choice of data structures, the heuristic crossover operator can be implemented to run as a linear function of the length of the structures [9]. This implies that, if E is the number of trials and N is the number of cities, our GA's for the TSP run with asymptotic complexity O(EN), the same as pure random search.

4. Experimental Results

This section describes some experiments with the adjacency representation and the heuristic crossover operator. For each experiment, N cities were randomly placed in a square Euclidean space. The initial population consisted of randomly generated tours. The selection method was based on the expected value model. The crossover rate was set at 50%, and there was no explicit mutation operator.

Figure 1 shows the results of a 50 city problem, Figure 2 shows a 100 city problem and Figure 3 shows a 200 city problem. Each Figure shows a representative tour from the initial population, the best tour obtained part way through the search, and the best tour obtained after the entire search, along with a randomly selected tour in the final population. It can be seen, especially in Figues 2 and 3, that good subtours tend to survive and to propagate. The figures also show that there is still a good deal of diversity in the final population.

Statistical techniques [2] allow us to estimate that the expected length of an optimal tour for experiment 1 is approximately 37.45. The optimal tour obtained by the GA differs from this expected optimum by about 25%. After an equal number of trials, random search produces a best tour of length 148.6, nearly 300% longer than the optimal tour. The optimal tour obtained in experiment 2 differs from the expected optimum by 16%. The optimal tour obtained in experiment 3 differs from the expected optimum by about 27%. These results are encouraging and suggest that further investigation of this approach is warranted.

Experiments show that GA's which use heuristic crossover but not selection perform better than random search but significantly worse than GA's which use both selection and heuristic crossover. That is, there appears to be a symbiotic relationship between the two levels of credit assignment performed by selection and heuristic crossover. We are currently working on clarifying the relationship between selection and the heuristic crossover operator.

5. Future Directions

This papers presents some preliminary observations and experiments. Many more questions about the TSP need to be investigated. Some interesting future projects include:

Combining GA's with other heuristics. In may be useful to heuristically choose the initial

164

population of tours. For example, the nearest neighbor algorithm can generate a set of relatively good tours when started from various initial cities. For very large problems, nearest neighbor can be approximated by choosing a random set of cities and taking the one closest to the current city. Heuristics could also be invoked at the end of the GA to do some local modifications to the tours in the final population. For example, the Figures shows many opportunities for improving the final tour by some local edge reversals.

Comparison with simulated annealing. Simulated annealing is another randomized heuristic algorithm which has been applied to very large (N > 1000) TSP's. From the published literature on simulated annealing [2,7], it appears that our results are at least competitive. A careful comparison of these two techniques would be very interesting.

Effects of GA parameters. There are several control parameters involved in any GA implementation, such as population size, crossover rate, etc. which may have an effect on the performance of the system. The proposed GA's are sufficiently different from previous GA's that it might be useful to investigate the effects of these parameters for the TSP.

Other combinatorial applications. How do the ideas developed thus far apply to combinatorial problems other than the TSP?

References

1. A. D. Bethke, *Genetic algorithms as function optimizers*, Ph. D. Thesis, Dept. Computer and Communication Sciences, Univ. of Michigan (1981).

2. E. Bonomi and J.-L. Lutton, "The N-city traveling salesman problem: statistical mechanics and the Metropolis Algorithm," *SIAM Review Vol. 26*(4), pp. 551-569 (Oct. 1984).

3. K. A. Dejong, *Analysis of the behavior of a class of genetic adaptive systems*, Ph. D. Thesis, Dept. Computer and Communication Sciences, Univ. of Michigan (1975).

4. K. A. Dejong, "Adaptive system design: a genetic approach," *IEEE Trans. Syst., Man, and Cyber. Vol. SMC-10*(9), pp. 556-574 (Sept 1980).

5. M. R. Garey and D. S. Johnson, *Computers and Intractability*, W. H. Freeman Co., San Fransisco (1979).

6. J. H. Holland, *Adaptation in Natural and Artificial Systems*, Univ. of Michigan Press, Ann Arbor (1975).

7. S. Kirkpatrick, C. D. Gelatt, and M. P. Vecchi, "Optimization by simulated annealing," *Science Vol. 220*(4598), pp. 671-680 (May 1983).

8. J. Pearl, *Heuristics*, Addison-Wesley, Menlo Park (1984).

9. B. J. Rosmaita, *Exodus: An extension of the the genetic algorithm to problems dealing with permutations*, M.S. Thesis, Computer Science Department, Vanderbilt University (Aug. 1985).

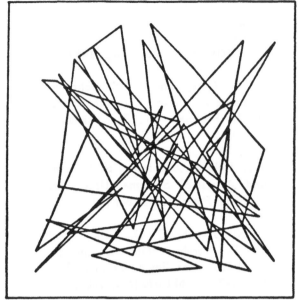

FIGURE 1a
50 CITIES
DISTANCE = 197.82
INITIAL POPULATION

FIGURE 1b
50 CITIES
DISTANCE = 64.76
GENERATION 38   1969 TRIALS

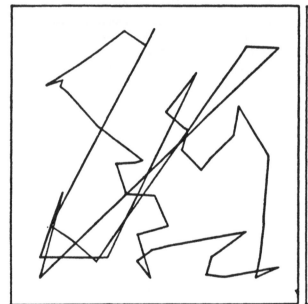

FIGURE 1c
50 CITIES
DISTANCE = 68.32
FINAL POPULATION

FIGURE 1d
50 CITIES
DISTANCE = 46.84
GENERATION 294   14686 TRIALS

Figure 1.

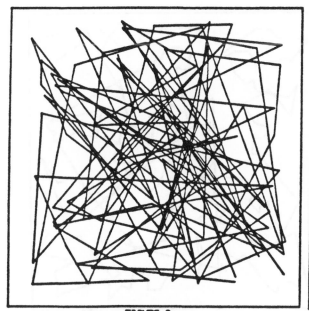

FIGURE 2a
100 CITIES
DISTANCE = 547.12
INITIAL POPULATION

FIGURE 2b
100 CITIES
DISTANCE = 118.47
GENERATION 125   6296 TRIALS

FIGURE 2c
100 CITIES
DISTANCE = 99.84
FINAL POPULATION

FIGURE 2d
100 CITIES
DISTANCE = 87.21
GENERATION 407   20330 TRIALS

Figure 2.

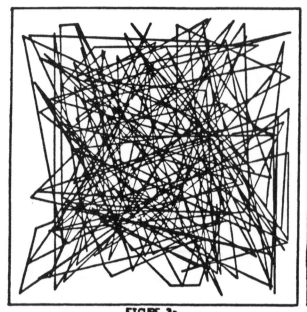

FIGURE 3a
200 CITIES
DISTANCE = 1475.68
INITIAL POPULATION

FIGURE 3b
200 CITIES
DISTANCE = 223.81
GENERATION 227   11373 TRIALS

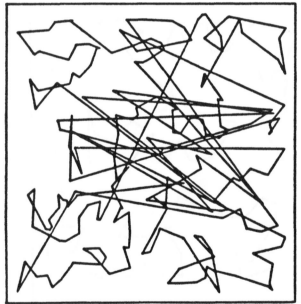

FIGURE 3c
200 CITIES
DISTANCE = 351.22
FINAL POPULATION

FIGURE 3d
200 CITIES
DISTANCE = 203.46
GENERATION 493   24596 TRIALS

Figure 3.

Genetic Algorithms: A 10 Year Perspective

Kenneth De Jong
George Mason University
Fairfax, VA 22030

1. Introduction

In 1975 Holland's book, *Adaptation in Natural and Artificial Systems*, was published and provided a summary of the work which Holland and his students had been pursuing for some time. An important theme in this wide ranging study of the properties of adaptive systems was that adaptation can be usefully modeled as a form of search through a space of structural changes which one might make to a complex system in an attempt to "improve" its behavioral characteristics. This gave rise to a methodology for studying existing (natural) adaptive systems and designing (artificial) adaptive systems which focused on answering key questions such as: What are the legal structural changes one is allowed to make? How is that space searched in an attempt to identify structural changes which improve behavior? How does one ascertain that resulting behavioral changes are, in fact, an improvement?

As an example of the merit of this approach, Holland specified the architecture for and provided a theoretical analysis of a class of adaptive systems in which the structural modification space is represented by strings of symbols chosen from some alphabet and the searching of this representation space is accomplished by an unusual procedure called a genetic algorithm. I think it is fair to say, at this point in time, that the careful definition and theoretic analysis of these genetic algorithms (GAs) was and continues to be one of the major contributions of this effort. In the intervening ten years, a good deal of interest and activity has resulted in important new insights into GAs and their potential applications, culminating in this conference.

Unfortunately, as is the case in many novel areas of research, it has been difficult to find a forum in the existing journal/conference structure for reporting the wide ranging activities which have resulted from Holland's provocative ideas. With only a few exceptions, much of this work has been disseminated via unpublished Master's and Ph.D. theses, personal communications, and presentations at a series of informal summer workshops.

I am pleased to report that this situation is changing for the better. In addition to growing institutional support for research in this area, the renewed interest in machine learning in the AI community as well as the continued interest in robust, flexible problem solving strategies in many different contexts has led to a dramatic increase in interest in GAs during the last few years. There remains, however, a fairly serious gap in the coverage of GA research activities since 1975. Those who are new to the area find it difficult to ascertain who has been doing what and frequently get involved unnecessarily in rediscovering various aspects of undocumented "wisdom" regarding the implementation and application of GAs. This conference in general and this paper in particular represent attempts to remedy such perceived gaps, to suggest open research issues, and to identify potential application areas. The following sections summarize my own personal perspective on the current state of the art in this field.

2. Conceptual and Perceptual Issues

Most algorithms are developed with a purpose in mind such as sorting, memory management, tree traversal, etc. Genetic algorithms, however, represent a highly idealized model of a natural process and as such can be legitimately viewed as a simulation at a very high level of abstraction. This tends to raise some conceptual and perceptual difficulties when trying to understand exactly what GAs do and how they might be used.

Much of the early GA research, in an attempt to simplify an already complicated situation, focused on understanding how GAs behaved when the structure space to be searched was an N-dimensional space of numerical parameters (corresponding to independently settable dials on a control panel) and the behavior of the system under the new control settings (the fitness measure) was ascertained by simply computing a memoryless function whose arguments were the new control settings. By carefully choosing functions which presented a variety of well understood payoff surfaces, a great deal of insight was obtained regarding how GAs distribute trials in such spaces in response to the feedback obtained from earlier trials. This gave rise to a very natural question: Do GAs provide a new and important technique for solving global function optimization problems? A good deal of research [DeJong75, Brindle80, Bethke81] has and continues to be done in this area with impressive results.

However, because of this historical focus and emphasis on function optimization applications, it is easy to fall into the trap of perceiving GAs *themselves* as optimization algorithms and then being surprised and/or disappointed when they fail to find an "obvious" optimum in a particular search space. My suggestion for avoiding this perceptual trap is to think of GAs as a (highly idealized) simulation of a natural process and as such they embody the goals and purpose (if any) of that natural process. I'm not sure if anyone is up to the task of defining the goals and purpose of evolutionary systems; however, I think it's fair to say that such systems are *not* generally perceived as function optimizers.

The question that remains, then, is how can one characterize what GAs do in a way which is useful for understanding how they might be best applied to difficult areas such as global function optimization, machine learning, NP-hard problems, machine vision, etc. I believe we still have a long way to go in this area. I have attempted to summarize recent advances as well as identify some open issues in the next section. To my mind the best perspective currently available as to what GAs do is

Holland's characterization of them as simultaneously solving a large number of K-armed bandit problems. (If you haven't read it or didn't understand it, you should make an effort to do so.) Although this characterization leaves many unanswered questions, armed with this viewpoint, one shouldn't be surprised that: 1) the best individual encountered so far may not even survive into the next generation, 2) that the population itself seldom converges to a global (or even local) optima, or 3) that the ability of GAs to produce a steady stream of offspring that are *better than any seen so far* can vary from quite impressive to dismal.

At the risk of summarizing the obvious, it is important to realize that GAs have properties of their own independent of the application area, and the key to a successful application (including global function optimization) is to understand and exploit these properties.

3. Representation Issues

The strongest hyperplane analysis results assume that GAs use a very specific form of selection, crossover, and mutation to search a space of fixed length binary strings. In order to take advantage of the power of GAs *as analyzed*, the space to be searched in a particular application must be mapped onto a representation space of this form. Depending on the application, selecting an appropriate mapping can range from a trivial activity to a highly creative one. There is now sufficient experience to begin to characterize search spaces with respect to choosing a representation mapping. The following is an attempt to do so.

3.1. Searching Parameter Spaces

Typically, the simplest way to make a complex process more flexible (adaptive) is to identify a fixed set of parameters which can be altered to improve behavior. The obvious mapping is to think of each of the N parameters as a genes and assign each a gene (string) position. If we then choose for each parameter a set of unique symbols representing the legal values of that parameter, we have a very intuitive internal representation as strings of length N. Crossover occurs between symbol boundaries and produces "legal" offspring, and mutation when applied to position i selects a new symbol from the legal symbol set for that position. There is both theoretical and experimental evidence to suggest that such direct intuitive mappings are appropriate when the number of legal values a parameter may take on is quite small (ideally, 2) and inappropriate when they deviate much from the ideal [Holland75].

Although there are many interesting problems which permit such direct mappings (e.g., feature spaces, certain NP-hard problems), most parameter modification problems do not. An obvious solution is to map each of the N symbol sets onto a set of fixed-length binary strings, concatenate the results, and apply GAs to this representation space. While it is easy to demonstrate a dramatic improvement in the behavior of GAs in switching from a short length, high cardinality representation of a problem to a longer, but lower cardinality representation, there are a several issues which arise for which we do not have good answers. Frequently the cardinality of a symbol set is not a power of 2, requiring rounding up to the next power of 2 and implying the symbol map is *into* but not *onto* the set of binary strings. In so doing, the size of the representation space can be increased (in the worst case) by a factor of 2^N over the original search space. Since crossover and mutation will invariably produce some of these unassigned strings, there are any number of ways to handle this including discarding such strings as illegal, assigning such strings low payoff, or mapping such strings redundantly into the symbol set. Each of these approaches has been tried at various times with no clear indication (either experimentally or theoretically) of the overhead incurred by such rounding or whether one approach is consistently better than another.

Frequently the application permits enough flexibility in defining the original search space so that the set of legal values each parameter can take on can easily and naturally be powers of 2 (e.g., most function optimization problems) so that rounding up issues are perceived as critical. There remains, however, the problem of selecting which of the M! ways M objects can be mapped onto another set of M objects in order to generate binary representations. This issue came up early in the function optimization studies in that when presented with certain relatively simple continuous surfaces GAs appeared to "lack the killer instinct" in the sense that they would quickly find near-optimal points, but fail to press on to better points near by. Further analysis indicated that such behavior was generally caused by artificial "representation boundaries" introduced by mapping the original space onto a binary representation space in such a way that "near-by-ness" had not been preserved. Hence, at a representation boundary, a small change in the value of a parameter is achieved only by a radical change in the binary representation of that parameter value. Since crossover and mutation are operating at the bit level, only very low probability sequences of events could "bump" the search over such boundaries. Experiments with alternative encodings such as gray codes yielded clearly identifiable improvements in cases where representation boundaries appeared to be a problem, but gave mixed results in others [Brindle80, Bethke81, ..]. Another suggestion for which there are no definite results is to redefine mutation so that it works at the parameter level, guaranteeing that at any point in time each parameter value is equally likely to be generated. The argument against such an approach is the disruptive effect such an operator would have on the proper allocation of trials to hyperplanes at the bit level.

As a consequence, an important open question is a better understanding of exactly what has to be preserved when choosing a mapping and how to find mappings with the desired properties. The only hints and suggestions along these lines that I am aware of are Bethke's use of Walsh transforms to characterize when representation spaces are "GA hard" [Bethke81]. Any new results in this area would greatly improve our understanding and use of GAs.

3.2. Adaptive Representations

Since there may not be sufficient *a priori* insight to select an appropriate representation, an alternative approach which has been discussed but for which there is little theoretical or experimental insight is to allow GAs *themselves* to select the mapping as part of the adaptive process. One strategy involves including extra "tag bits" with each individual which identifies the mapping to be used. An interesting issue here is whether GAs should be modified to be aware of such tags bits (for example, by only applying crossover to parents with identical mappings) or whether GAs should manipulate the tags bits in the usual way as undistinguished members of a longer binary string. In the former case, this introduces the idea of subpopulations (species) for which there is considerable support in natural systems but for which there are no analytic results. In the latter case, the presumed usefulness of binary strings inherited from one (and possibly both) parents can be lost because they are interpreted in a totally different way in an offspring unless the parents had identical tag bits and mutation left them unchanged.

Holland raised similar issues while analyzing the disruptive effects of crossover on co-adapted sets of alleles which, because of the particular representation chosen, happened to be far apart [Holland75]. His suggestion was to introduce the inversion operator as a mechanism for changing the physical location of genes without changing their functional interpretation. As above, left unresolved were issues such as whether there should only be a few inversion patterns (species) present in a population with mating (crossover) occurring only within species or whether crossover should be modified to allow offspring to inherit an inversion pattern from one parent but gene values from both. Early experimental work [Franz72, DeJong75] generated little evidence of any significant improvement due to introducing inversion in a function optimization context; however, inversion proved to be effective in later work using GAs to search spaces of production system programs [Smith80].

3.3. Context Sensitive Values

A related but more fundamental problem arises when the application area has the property that the legal values for one parameter are *context sensitive* in that they depend on which values have been chosen at other positions. While it is frequently convenient and natural to view such problems as defining parameter spaces to be searched, violating the assumption that values can be selected independently can have dramatic effects on the performance of GAs. A simple example of this occurs if we try to represent the unit circle with Cartesian coordinates mapped onto fixed-length strings. GAs, by independently choosing symbols at each position, will distribute trials over the unit square. The usual "fix" is to define the payoff outside the unit circle to be exceptionally low (a penalty function) and let the GAs "learn" to keep new trials inside the desired region. Suppose, however, we generalize the problem to that of representing an N-dimensional hypersphere using Cartesian coordinates. If GAs distribute their trials over the enclosing hypercube, as N gets large, the volume of the hypersphere becomes vanishingly small relative to the hypercube and the search process becomes hopelessly bogged down on a surface which appears to be uniformly bad almost everywhere. In this case, of course, it doesn't take much insight to suggest a switch to polar coordinates. However, there are other cases in which alternate representations are not so easy to find.

My favorite example of this is the Traveling Salesman Problem (TSP), and I am delighted to see that it is well represented at this conference. I continue to believe that it captures in a simple, elegant way many of the open GA issues. A good deal of thought and discussion has gone into the problem of representing TSPs in a form amenable to GAs with very little success to this point. Since the problem involves visiting each of N cities exactly once while minimizing the total distance of a tour, the most natural way to represent candidate solutions is to list in order the cities visited. Obviously, even though this representation can be viewed as N parameters specifying the Ith city to be visited, it is strongly context sensitive in that once a city symbol is used, it cannot be re-used in another position. Of course, one can always permit the GAs to construct illegal tours via crossover and mutation and assign them a very low payoff. Unfortunately, just as with hyperspheres, the space of interest here (the set of all *permutations* of N symbols) becomes a vanishingly small fraction of the the set of all *combinations* as N increases. There have been many alternative representations invented and explored, but to my knowledge none represent the set of permutations in an efficient, context free way.

The alternative to finding a representation which fits with the standard versions of crossover and mutation is to change the definition of crossover and mutation to fit the representation. Inventing new mutation operators is not too difficult in this case, the most natural being low order permutation operators. Crossover requires a bit more creativity and usually involves taking a partial tours from one parent and splicing in whatever is legally possible from the second parent. The results to date from this approach have not been any more encouraging than the previous ones using the standard versions of crossover and mutation on inadequate representations. The problem in this case is that, by altering the genetic operators, we have altered the way in which GAs distribute trials and the fundamental theorems regarding efficient parallel search need to be re-proved.

So we find ourselves "caught between a rock and a hard place" with few places to turn. I don't claim to have the answer either, but there are several observations which would seem to provide some hints. TSP problems fall into an equivalence class of problems called NP-complete because there are no known polynomial-time solutions for any member of the class and if one were found there are polynomial-time transformations permitting all other members to be solved in polynomial time. The Boolean Satisfiability Problem (BSP) is a member of this class and involves finding truth value assignments to N boolean variables in such a way as to make an arbitrary given boolean expression of these N variables true. The most natural representation for BSPs is precisely what is needed for use with GAs, namely a binary string of length N. Crossover and mutation work precisely as intended and problems of surprising size can be solved. (Unfortunately, there isn't much interest here in nearly correct assignments!) What we have then are two problems which are known to be equivalent in the NP-hard sense, but are quite different in a GA-hard sense.

The difference seems to hinge on a sort of duality relationship between the two problems. Fitness for BSPs is defined purely in terms of the values of the symbols and not their relative positions in the string. This maps well onto our notion of

hyperplane and in these situations crossover and mutation are effective mechanisms for homing in on good value combinations. On the other hand, TSP fitness is defined purely in terms of the order of valueless genes which represents being in city n. Here inversion seems most natural with crossover and mutation inappropriate in their usual form. What seems to be needed is a definition of a hyperplane in this dual space. Unfortunately, our notions of hyperplanes are so tightly bound to spaces represented by a fixed number of independent axes that it's hard to conceive of alternate definitions. With an appropriate definition there would be a much clearer view of the duals to crossover and mutation, and hopefully a dual set of analytic results.

3.4. Context Sensitive Interpretations

Another form of context sensitivity can arise and cause difficulty when the same value of a particular parameter has different interpretations depending on the values of other parameters. We have already seen how this can occur when attempting to select representations adaptively. Another nice example arises in attempting to escape from the context sensitive value representations of TSPs. One could imagine an N parameter representation in which the first parameter specified which of the N cities should be visited first. Having deleted that city from our list, the second parameter always takes on a value in the range 1...N-1, specifying by position on our list which of the remaining cities is to be visited second, and so on. Values for each of the parameters can now be independently selected and crossover and mutation always produce legal tours. However, the performance of GAs on this representation is not significantly better than the previous ones. The difficulty appears to be that gene values to the right of a crossover point or a mutation are interpreted quite differently (i.e., specify totally different subtours) in an offspring than in the parent, violating the concept of minimal disruption of "building block" formation. What seems to be needed is a representation which allows good subtours (co-adapted sets) to form and be passed on in combination with other subtours, forming better tours, and so on. With the traditional definition of a hyperplane, this seems to rule out context sensitive interpretations as bad representations. I am unaware of any alternatives other than the hope that perhaps a more general perspective on hyperplanes will clarify these issues.

3.5. Varying Length Representations

So far we have been discussing issues which appear in the context of searching parameter spaces. There are, of course, many other (generally more complex) kinds of spaces which represent the set of permissible structural changes to an adaptive process. In some cases strings are still a natural representation, but there may be no notion of a fixed length. A good example are strings which specify structural changes via "genes" which represent actions to be taken. One string may consist of only a few actions while others require many. If we wish to use standard GAs, the simplest (but somewhat inefficient) approach is to assume some reasonable upper bound on the length, throw in a "no-op" action, and require all strings to be maximum length. Alternatively, crossover can be easily generalized to produce offspring whose length is different (in general) from either parent by choosing independent crossover points in each parent.

However, it is important to note that neither approach is sufficient to guarantee good GA performance on varying string length spaces. To understand why requires asking what the hyperplanes are in this context. Both Holland [Holland75] and Smith [Smith80] discuss the issues. I will not repeat the discussions here, but just note that there is considerable evidence that a sufficient condition for good GA performance is that the genes express their actions in a position independent way.

3.6. Non-String Representations

What should one do when elements in the space to be searched are most naturally represented by more complex data structures such as arrays, trees, digraphs, etc. Should one attempt to "linearize" them into a string representation or are there ways to creatively redefine crossover and mutation to work directly on such structures. I am unaware of any progress in this area. However, the issues appear to be reasonably clear. Any linear representations will have

to satisfy the properties discussed in the preceding sections in order to achieve efficient GA search. Similarly, any attempts to modify crossover and mutation will require analogous hyperplane analysis results to guarantee reasonable performance.

3.7. Production System Spaces

One of the most intellectually pleasing ways to effect changes in the behavior of a complex process is to modify its knowledge base. There has been a good deal of research within the AI community regarding appropriate ways to represent knowledge. Production rules are frequently chosen when learning is involved [Waterman70, Newell77, Buchanan78]. The GA community has also maintained a long standing interest in production system architectures because of their amenability for use with GAs [Holland75, Holland78, Smith80, Booker82]. From my perspective there are currently two main approaches to searching production system rule spaces with GAs.

The first is typified by the classifier systems developed initially by Holland [Holland78] and Booker [Booker82]. Here individuals in the population represent single production rules (typically fixed length) and the current population represents the entire set of rules governing the behavior of the adaptive process. GAs play a subservient role within a larger cognitive model and are invoked intermittently to produce new rules which replace existing rules in the population.

The alternate approach is represented by the LS-1 system developed by Smith [Smith80]. Individuals represent entire rule sets to be plugged into the knowledge base and evaluated. The next generation of rule sets is produced in the usual way by applying genetic operators to existing rule sets.

Both approaches have produced encouraging results in quite different contexts. There is not enough experience, however, to understand precisely the strengths, weaknesses, and tradeoffs involved in either of the approaches. My guess is that the classifier approach will prove to be most useful in an on-line, real-time environment in which radical changes in behavior cannot be tolerated whereas the LS-1 approach will be best suited for off-line environments in which more leisurely exploration and more radical behavioral changes are acceptable.

4. Fitness Functions

In addition to choosing an appropriate representation on which to apply GAs, careful thought must be given to the characteristics of the payoff function used to provide feedback regarding an individual's fitness to produce offspring. The wealth of data from GA function optimization studies simultaneously show a general robustness in performance over widely varying classes of functions and intermittent dismal results. This has lead to several informal characterizations of the kinds of surfaces which are GA-hard. Surfaces which are flat almost everywhere except for an occasional spike present difficult search problems for any approach including GAs. The intuitive explanation is that, since there is (essentially) no differential payoff among the competing hyperplanes, such peaks will be found only by chance. Unfortunately, it is not all that difficult to inadvertently construct one in applications like the hypersphere and BSP examples discussed earlier.

This immediately suggests another way to fool GAs: put misleading information in the hyperplanes. Fortunately, this is much more difficult to do because of the simultaneous sampling of many different hyperplane partition elements. Bethke [Bethke81] has a nice discussion of this using Walsh transforms to characterize GA-hard functions. However, much more work needs to be done in this area.

It should be also noted that it is quite easy to incorrectly blame GAs for poor performance when the fault in fact lies elsewhere. One classic case of this arises when using GAs to improve the performance of a complex process for which no payoff function is given. Since one has to be constructed, care must be taken to verify that high payoff values as seen by GAs corresponds to good behavior as observed by watching the complex process itself. Another case arises when numeric parameter spaces are being searched. Since there is typically some freedom in how finely to discretize a parameter range, choosing too coarse a discretization factor may inadvertently leave out optimal points in the representation space being searched by GAs and then blame the GAs for not finding them!

Until recently, most GA research and applications involved payoff functions which return a single (scalar) payoff value. There are situations in which it is more natural to have the payoff function return a vector of values representing, for example, scores on non-commensurate aspects of performance. Rather than insisting that an artificial function be created which combines such scores into a single payoff value, it would be preferable to have GAs work directly with multi-valued payoffs. Schaffer [Schaffer85] has explored this possibility recently and has obtained promising results.

5. Genetic Operators

There certainly is nothing sacred about the traditional operators defined and analyzed by Holland. What is important is that we have criteria from Holland's hyperplane analysis which operators should meet. If changes are made to existing operators or new ones are introduced, it is important to verify that they aren't overly disruptive of the process of distribution of trials according to payoff and that they encourage the formation of building blocks. There are still some interesting open questions along these lines with respect to rather modest variations of the standard operators.

It is pretty much standard procedure now to view crossover as applying to circular strings and selecting two crossover points, the beginning and the end of the segment provided by the second parent. This modification is well supported both theoretically and experimentally. What happens if we continue along this vein and select two segments from the second parent (via four crossover points)? Is this helpful or too disruptive? The answers are pretty clearly negative by the time we have increased the number of crossover points to the extent that an offspring's gene values are randomly selected from its parents values. Perhaps the number of crossover points should be a function of the length of the strings involved. Applying the traditional crossover to strings with thousands of genes (which is currently being done) seems to be intuitively more disruptive than one with four or six crossover points. If so, where does the law of diminishing returns set in?

The role of mutation as a background operator which introduces new allele values is fairly well understood and accepted in the abstract. As discussed earlier, problems can arise from our choice of representation in which mutation (and crossover) are operating at the bit level, but our interpretation of the search space is at a higher level. This can lead to a frequently tried but rarely successful strategy of increasing the mutation rate to improve GA performance. A better approach in such situations is to think in terms of both higher and lower level versions of the genetic operators. Both Holland [Holland75] and Smith [Smith80] discuss this, but much more work needs to be done.

6. Selection

The technique of selecting parents for reproduction with a frequency proportional to observed fitness has strong theoretical justification and considerable empirical support. However, there are occasions when this process seems to break down when implementing GAs with finite populations. This has come to be known as "the scaling problem" and can occur in a number of ways. If a highly fit individual is encountered early in the search process among mediocre peers, selection will give it such strong preference that it can dominate the population in a few generations and cause premature convergence. Similarly, late in the search process the population can be legitimately dominated by members with very high payoffs which differ on an absolute scale, but when normalized to produce expected number of offspring are equivalent out to the third or fourth decimal place. The effect is that essentially every parent contributes equally to subsequent populations in spite of fitness differences.

There have been a number of proposed solutions including the introduction of scaling factors and crowding factors [DeJong75], and selection by rank [Wetzel83, Shaffer85]. However, I think it is fair to say that a general solution still eludes us.

7. GA Parameters

One of the observations people are quick to make is that GAs are themselves complex processes which appear to have a set of parameters (crossover rate, mutation rate, population size, etc.) which could be tuned to improve performance. There is considerable empirical support for the statement that within reasonable ranges the values of such parameters are not all that critical [DeJong75, Grefenstette85]. As a consequence most GA applications work with fixed "accepted" parameter values. However, there is also evidence to suggest that additional performance improvements could be obtained if such parameter values could be dynamically modified. The difficulty is in deciding when and how to effect such changes. Should we have a two-level GA complex with the top level GA actively searching the parameter space of the lower level GA and trying out new parameter combinations? Are there simpler signals such as allele loss which should trigger parameter changes? Unfortunately, the existing theory gives little guidance here.

8. Conclusion

In rereading the previous sections, I became a little concerned that the reader might infer a strong negative tone from this long list of problems and open issues in GA research. Nothing could be further from my intent. I am enthusiastic about the potential which GAs hold and am actively involved in GA research and applications. It is that enthusiasm which generated this paper and this conference. I hope the result is that the next time we get together my list will be considerably shorter (or at least different)!

References

[Bethke80] Bethke, A., "Genetic Algorithms as Function Optimizers", Doctoral Thesis, CCS Department, University of Michigan, 1981.

[Booker82] Booker, L. B., "Intelligent Behavior as an Adaptation to the Task Environment", Doctoral Thesis, CCS Department, University of Michigan, 1982.

[Brindle80] Brindle, A., "Genetic Algorithms for Function Optimization", Doctoral Thesis, Department of Computing Science, University of Alberta, 1980.

[Buchanan78] Buchanan, B., Mitchell, T.M., "Model-Directed Learning of Production Rules", in *Pattern-Directed Inference Systems,* eds. Waterman and Hayes-Roth, Academic Press, 1978.

[DeJong75] De Jong, K., "The Analysis of the Behavior of a Class of Genetic Adaptive Systems", Doctoral Thesis, CCS Department, University of Michigan, 1975.

[DeJong80a] De Jong, K., "A Genetic-based Global Function Optimization Technique", TR 80-2, Department of Computer Science, University of Pittsburgh, 1980.

[DeJong80b] DeJong, K., "Adaptive System Design: A Genetic Approach", IEEE Trans. on Systems, Man and Cybernetics, 10,9, Sept. 1980.

[DeJong81] De Jong, K. and Smith, T., "Genetic Algorithms Applied to Information Driven Models of US Migration Patterns", 12th Annual Pittsburgh Conf. on Modelling and Simulation, April 1981.

[Frantz72] Frantz, D. R., "Non-linearities in Genetic Search", Doctoral Thesis, CCS Department, University of Michigan, 1972.

[Grefenstette85] Grefenstette, J., "Genetic Algorithms for Multilevel Adaptive Systems", to appear in IEEE Trans. on Systems, Man and Cybernetics.

[Hedrick76] Hedrick, C.L., "Learning Production Systems from Examples", Artificial Intelligence, Vol. 7, 1976.

[Holland75] J. H. Holland, *Adaptation in Natural and Artificial Systems.* University of Michigan Press, 1975.

[Holland78] Holland, J.H., Reitman, J., "Cognitive Systems Based on Adaptive Algorithms", in *Pattern-Directed Inference Systems,* eds. Waterman and Hayes-Roth, Academic Press, 1978.

[Newell77] Newell, A., "Knowledge Representation Aspects of Production Systems", *Proc. 5th IJCAI,* 1977.

[Schaffer85] Schaffer, J. D., "Multiple Objective Optimization with Vector Evaluated Genetic Algorithms", to appear in Proc. Int'l Conf. on Genetic Algorithms and their Applications, July 1985.

[Smith80] Smith, S. F., "A Learning System Based on Genetic Adaptive Algorithms", Doctoral Thesis, Department of Computer Science, University of Pittsburgh, 1980.

[Smith83] Smith, S. F., "Flexible Learning of Problem Solving Heuristics Through Adaptive Search", Proc. 8th IJCAI, August 1983.

[Wetzel83] Wetzel, A., "Evaluation of the Effectiveness of Genetic Algorithms to Combinatorial Optimization", Doctoral Thesis, Department of Library and Information Science, University of Pittsburgh, 1983.

Classifier System with Long-term Memory in Machine Learning

Hayong Zhou
Vanderbilt University

ABSTRACT

This paper discusses the advantages of classifier systems with long-term memory and includes a description of the basic structure of such a system. The learning strategy used here is twofold one. First, an analogical learning strategy is employed to inject the appropriate knowledge into the population. Second, a production system with a GA-based learning component is invoked to perform subsequent learning. The proposed system has one overall objective: It seeks to increase the efficiency and power of the learning system over a long period of time of use.

1. Introduction

A genetic algorithm (GA) is a problem-solving and non-deterministic search algorithm first introduced by Holland in 1975[3]. It has been shown, theoretically and empirically, that GAs are robust and effective in various task domains, even in the presence of difficulties such as noise, high-dimensionality, multimodality and discontinuity[7].

The outgrowth of the continuing research in this area evolved into a message-passing, rule-based production system called classifier system[4]. A classifier system is a learning system in which many classifiers are active simultaneously. A classifier is a pattern sensitive element with condition/action form. Each condition specifies the set of messages satisfying it, and each action specifies the message to be sent when its condition part is satisfied. In short, a classifier system manipulates knowledge structures (KSs) in response to performance via a genetic algorithm. It provides a framework for cognitive simulation[2].

Several published classifier systems which incorporate transfer of learning knowledge from one task to another have been developed: In 1978, Holland and Reitman designed the first classifier system called CS-1 tested on maze problems. An experiment was conducted to demonstrate transfer of learning from a small maze problem to a large but similar one[4]. The experimental result showed that CS-1 was able to solve the large maze problem much faster when initially supplied with some learned knowledge structures. In 1982, Booker did in-depth simulation study of classifier systems as cognitive models[2]. He performed several experiments to demonstrate the effects of prior knowledge structures on learning in new situation. For "positive transfer"(transfer of knowledge for solving similar tasks), his results were very encouraging.

Before proceeding any further, the "reversal learning task" needs to be described: Schrier[6] trained a monkey on a reversal learning task. Reward and punishment were reversed repeatedly while keeping the input information to the monkey unchanged. Performance of this monkey was inefficient at the outset, but, eventually, each new reversal could be learned with a single trial.

In order to test the learning ability of

178

classifier systems, Booker ran his system on the reversal learning task. Surprisingly, the resulting performance was inconclusive. The reasons, according to Booker, are that "the emphasis on recency and short-term memory in the system is too great" because "by the time the organism had reached criterion on a given reversal, the classifiers learned during the previous reversal were likely to have been deleted - that is, become "extinct" due to the drastic change in the environment"[2]. In 1984, Schaffer completed the LS-2 designed for the pattern discrimination task domain[5]. He also gave the reversal learning task to his system. The results obtained so far are not encouraging either(private communication).

In sum, efforts to build powerful classifier systems have met with impressive success over the past. The attempts to transfer learned knowledge for solving similar tasks, though manaully, have been shown to be useful and effective. However, the failures in solving the reversal learning task pose a question: Is there any way that classifier systems can keep knowledge which is useful but irrelevant to the current situation intact in order to increase the efficiency and power of their learning ability? To answer this question, this paper proceeds from a general need for having a long-term memory to a proposed prototype in the following sections.

2. Motivation for the design of classifier system with long-term memory(CSLM)

We begin this section with several assumptions which have been associated with traditional classifier systems.

- The domain of learning is concerned with a single task.
- The changes in environments are slight, smooth and gradual.
- The efficiency for solving similar tasks in a long run is not important.

If task domain satisfies these assumptions, it would be unnecessary to augment a classifier system with long-term memory. However, an ideal learning system should be able to switch its attention as needed while still preserving the most useful knowledge gained in the past no matter how its environment has been changed. By doing so, the system would increase its efficiency and power over time and improve its learning ability as the number of learned tasks grows.

In short, the main concern of this paper is to investigate how to accumulate and preserve knowledge not only within a task, but also among tasks. It has been shown empirically that the size of a population should be chosen around 50(number of knowledge structures) in order to maximize computational efficiency[8]. In practice, most of classifier systems never use a population larger than 200. For such small knowledge pools, it is hard to imagine that a set of generalized knowledge structures could be constructed, for example, suitable for many pattern discrimination tasks. A short-term memory, i.e. the population in a classifier system, can not be expected to meet the challenges imposed by drastic environmental changes. Each knowledge structure in a population is evaluated by the Critic designed for the current task. It is very difficult, if not impossible, to preserve those knowledge structures which were perfect for some previous tasks but not suitable for the current situation. We see this as a serious weakness of the current model and as the major motivation for the design of a classifier system with long-term memory (CSLM).

3. Overall description of CSLM

In this section, an overall organization of CSLM is outlined. The description is based on the following diagram and is intended to be instructive rather than specific. In figure 1, an understanding of the basics of classifier systems has been assumed. It is well described in [2,4,5].

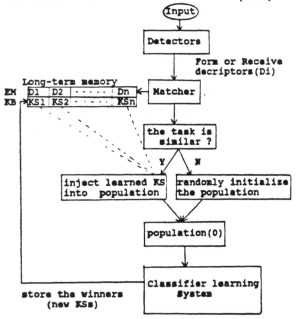

Figure 1   The paradigm of CSLM

In simplest terms, we can visualise the main components of CSLM as follows:

- **Descriptors:** Descriptors serve as indices to learned knowledge structures. The descriptors for various tasks could be very general. In fact, complete and precise descriptor for a task is neither necessary nor realistic. In practice, the descriptors might use a low level language(a string of bits) or a high level language(alphabet) to express main characteristics of tasks. They may be produced automatically from incoming tasks, or supplied by users.

- **Matcher:** The Matcher(a procedure) performs two functions: matching descriptors and initializing a population. We discuss them together here. Matching the descriptor of a incoming task with that of solved tasks in a long-term memory might

have one of the following three outcomes:

1. Exact matching.
   The next step is to bring the learned Knowledge structures into the population. Heuristic initialization of the population is done.

2. Partial matching.
   One similar task can be found. The similarity between the incoming task and the stored ones indicates that there might exist some useful building blocks in the stored knowledge structures which, hopefully, can provide a promising direction to start with. Thus the search space would be pruned and the computational effort might be reduced.

3. No matching.
   It tells us that no previous experience regarding the incoming task is known, or possibly has been forgotten. In this case the CSLM has to start from scratch, no worse than current classifier systems.

Long-term memory: The long-term memory consists of two separated memories called Episodic Memory(EM) and Knowledge Base(KB) respectively.

The EM stores all descriptors for previous tasks. Each descriptor has one pointer pointing to its corresponding KSs in the KB. The content of the EM may be considered as the indices for accumulated knowledge structures.

The KB preserves learned KSs.

Whenever a task has been solved, the set of solutions are stored in the long-term memory along with the associated pointer.

One of the basic learning strategies

employed in CSLM is "learning by analogy" which appears to be a central inference method in human cognition and promises to be a powerful mechanism in machine learning . Learning by analogy consists of two phases. The first phase is called the "reminding phase" which identifies the similarity between an incoming task and the problems observed or solved before. The second phase involves the transfer of appropriate knowledge obtained in the past into the new situation. Carbonell pointed out the importance of learning by analogy:" In general, transfer of experience among related problems appears to be theoretically significant phenomenon as well as a practical necessity in acquiring the task - dependent expertise necessary to solve more complex real world problems"[1].

The approach used in CSLM is to form descriptors derived from the detector array to categorize tasks.   In the reminding process, similarity could be determined by matching these stored descriptors in a long-term memory with the descriptor derived from an incoming task. In the next phase of analogical problem solving, the related knowledge structures, if any, would be brought into the population.   Notice that to inject these learned KSs into a population is not the end of our story. Instead, it should be viewed as providing strong guidance for future search. The genetic algorithm will manipulate these useful building blocks and transform them into a form that would be appropriate for the current task. In the next phase, the classifier system is invoked to perform the subsequent learning which will not been detailed here.

4. Solving the reversal learning task in CSLM

First of all, we need to emphasize that the interestingness of the reversal learning task is not only because it represents a new class of learning tasks, but also, more importantly, it tests the learning ability of a system on how well it can preserve useful knowledge from radical changes in environments.

Let us see what will happen if a reversal learning task is given to CSLM.

Suppose that a CSLM has created a set of KSs for a given task and stored it along with its associated descriptor in a long-term memory, as shown in figure 2.a. When the second task with the same appearance but opposite meaning(reversal task) is given, the CSLM, as expected, is in the worst possible position to learn the new task since the Matcher procedure would have brought the learned KSs into the population. In this case, the learned KS would receive a low score and the classifier system would have to develop a new KS for the reversal task. However, after the CSLM has created two sets of KSs for each reversal, it can solve subsequent reversal learning tasks with a single trial.   As noted earlier, the generality of a descriptor for a task would guarantee the CSLM to recognize the tasks with the same or similar characteristics.   Thus the Matcher would be able to pull two sets of KSs out of the long-term memory based on the similarity measurement and inject them into the population. The initialized population is shown in figure 2.b. Therefore, the Critic would be able to choose the appropriate KS for each reversal.

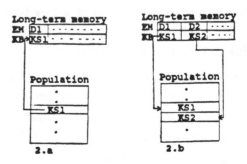

Figure 2

Another significance of this demonstration

is to show what happens if a set of bad knowledge structures has been used to initialize the population. The full power of genetic algorithms comes from the parallel nature of the search and the immunity to false peaks. Therefore, these injected KSs are only tentative, and as such are subject to testing. If some of them prove useless or misleading, they will die out in subsequent generations.

There is a further point worth noting: the portion of a population to be heuristically initialised should be judiciously decided so that the premature convergence could be avoided while still giving an opportunity for guiding future search.

5. Summary and Future Research

This paper has discussed the advantages of augmenting classifier systems with long-term memory and described a prototype of CSLM conceptually. The process of solving the reversal learning task was demonstrated as well. The driving force behind this paper is to extend the current model in order to deal with more complex task and make consistent progress even if environments have been drastically changed. Several difficulties which can be anticipated in the design of CSLM are mentioned here:

- How to extract descriptors from tasks with reasonable accuracy and effort while maintaining the delicate balance between generality and specificity?

- How to update the content of a long-term memory dynamically?

- How to best initialize a population?

In seeking the answer to these questions and to test the feasibility of the proposed ideas, a specific CSLM designed in the pattern discrimination domain is to be implemented. It is hoped that the experimental results will be available soon as an evidence of the improved learning ability of the proposed system.

Acknowledgements

The author would like to thank his advisor, Dr.John Grefenstette for his guidance, and Dr. David Schaffer for his encouragement during the development of this paper.

References

1. **Carbonell, J.G.** Learning By Analogy :Formulating And Generalizing Plans From Past Experience. Machine learning, p137-159, Tioga publishing Co.

2. **Booker, L.** Intelligent Behavior As An Adaptation To The Task Environment Ph.D. dissertation, The university of Michigan, 1982

3. **Holland, J.H.** Adaptation in Natural and Artificial System The university of Michigan press, 1975

4. **Holland, J.H. and Reitman, J.S.** "Cognitive Systems Based On Adaptive Algorithms, Pattern-directed inference system, 313-329, 1978

5. **Schaffer, J.D.** Some Experiments in Machine Learning Using Vector Evaluated Genetic Algorithm Ph.D. dissertation, Vanderbilt University.1984

6. **Schrier, A.M.** Transfer By Macaque Monkey Between Learning-set and Reperted-reversal Tasks. Percept. Mot. Skills, 23, 787-792

7. **Dejong, K.A.** Analysis of the Behavior of a Class of Genetic Adaptive Systems. Ph.D. dissertation,Univ. of Michigan, 1975

8. **Grefenstette, J.J.** Optimisation of Control Parameters for Genetic Algorithm, To appear in IEEE Trans. Sys. Man, Cybn,1985

A Representation for the Adaptive Generation of Simple Sequential Programs

Nichael Lynn Cramer

Texas Instruments Inc.

PO Box 226015,MS 238

Dallas, TX 75266

ABSTRACT

An adaptive system for generating short sequential computer functions is described. The created functions are written in the simple "number-string" language JB, and in TB, a modified version of JB with a tree-like structure. These languages have the feature that they can be used to represent well-formed, useful computer programs while still being amenable to suitably defined genetic operators. The system is used to produce two-input, single-output multiplication functions that are concise and well-defined. Future work, dealing with extensions to more complicated functions and generalizations of the techniques, is also discussed.

INTRODUCTION

The techniques of adaptive Genetic Algorithms [GAs][1] have been shown to be useful in many areas. Initially, these systems involved the adjusting of a fixed set of parameters in order to optimize the performance of a given algorithm[2]. Much work has been done toward the goal of evolving the algorithms *themselves*, particularly in Production System-like domains[1(chap8),3,4]. This paper discusses work towards developing a sequential programming language that is suitable for manipulation by GAs so as to permit the adaptive generation of simple computer functions from low-level computational primitives.

FUNCTIONAL REPRESENTATION

The scheme that we will follow is first to find a suitably powerful programming language, and then encode the programs in this language in such a way as to make them amenable to the standard Genetic Operators [GOs].

The basic language to be used is a variation of the algorithmic language PL having the following operators:

(:INC VAR) ;;add 1 to the variable VAR

(:ZERO VAR) ;;set the variable VAR to 0

(:LOOP VAR STAT) ;;perform the statement STAT VAR times

(:GOTO LAB) ;;jump to the statement with label LAB

Programs in PL consist of an arbitrary number of globally-scoped (positive) integer variables and statements containing operators of the above forms. Two simple example PL Programs are:

;;Set variable V0 to have the value of V1

(:ZERO V0)

(:LOOP V1 (:INC V0))

;;Multiply V3 to V4 and store the result in V5

(:ZERO V5)

183

(:LOOP V3 (:LOOP V4 (:INC V5)))

While **PL** can be shown to be Turing Equivalent [5], we will be interested in the language subset **PL**-{: $GOTO$}. This language subset has two useful properties: first, while it is not fully Turing Equivalent, it still comprises a powerful set of functions (specifically, the set of primitive recursive functions)[5] and second, programs written in **PL**-{: $GOTO$} are guaranteed to halt. Finally, we make two small extensions to the language. First, a :SET operator, which accepts two variables and sets the value of the first variable equal to that of the second. (As can be seen in in the examples above, this operation is trivially definable in **PL**-{: $GOTO$}; if so desired, it can be considered a macro or subroutine operator.) Secondly, we define a :BLOCK operator that accepts two statements as arguments and evaluates the two statements sequentially. (This is essentially just a grouping operation that has no effect on the overall structure of the language.)

Now, the encoded representation for our programs should have two characteristics:

(Goal 1) It should be amenable to the standard **GO**s.

(Goal 2) The representation should produce only well-formed programs, even when subjected to the **GO**s. While some representations, e.g. character-strings, might be well suited for the mechanisms of **GO**s, the random generation and/or altering of characters is not likely to produce, say, a useful **FORTRAN** program. Consequently, it is strongly desirable that the chosen representation be such that all such generated programs stay in the space of syntactically correct programs. Not all such generated programs would be useful (adapation would be expected to correct that); it is only important at this point that such programs be well formed.

This paper will consider lists of integers as a representation for these programs where the object the integer represents (variable, operator, etc,) is determined by the integer's position in the list. Clearly such a representation satisfies Goal 1 above, the standard **GO**s (Crossover, Mutation, Inversion) would be well defined on such a list. To satisfy Goal 2, we need to define a decoding of an arbitrary list into a well-formed program.

THE JB LANGUAGE

A first attempt at such a decoding is the language **JB**. The list of integers is first divided into statements of some length large enough for the longest statement size, (three in the present case). Any integers left over at the end of this list are ignored. The first of these statements is defined to be the Main Statement [MS] and the remaining N_{as} statements are the Auxiliary Statements [AS]. Syntactically, these statements are interpreted as follows:

(0 4 2) -> (:BLOCK AS_4 AS_2)

(1 6 0) -> (:LOOP V_6 AS_0)

(2 1 9) -> (:SET V_1 V_9)

(3 17 8) -> (:ZERO V_{17}) ;;the 8 is ignored

(4 0 5) -> (:INC V_0) ;;the 5 is ignored

Here the symbols of the forms V_n and AS_n represent, respectively, example Variables and Auxiliary Statements.

This body of statements is embedded in an environment containing N_{bv} body-variables (initialized to 0) and N_{iv} input-variables. At the end of the execution of the program, any of the $N_{vtot} = (N_{iv} + N_{bv})$ available variables can be returned as ouput.

The function is entered by executing the MS, which, typically, will call on one or more of the AS's. An example JB program would be:

(0 0 1 3 5 8 1 3 2 1 4 3 4 5 9 9 2)

This would be grouped into the following Statements:

(0 0 1) ;;main statement -> (:BLOCK AS_0 AS_1)
(3 5 8) ;;auxiliary statement 0 -> (:ZERO V_5)
(1 3 2) ;;auxiliary statement 1 -> (:LOOP V_3 AS_2)
(1 4 3) ;;auxiliary statement 2 -> (:LOOP V_4 AS_3)
(4 5 9) ;;auxiliary statement 3 -> (:INC V_5)

This is the same as the PL multiplication program above.

As can be seen, virtually (see below) any list (of sufficient length) of integers chosen from the range $[0, N_{rand}-1]$ can be used to generate a well-formed JB program. Where $N_{rand} = N_{vtot} * N_{as} * N_{op}$ (N_{op} is the total number of operator types). A particular language object (variable, AS, operator-type) needed for the program can then be extracted from a given integer in the list by taking the modulus of that integer with respect to the respective number above. This ensures random selection over all syntactic types. Two problems arise from this straight forward use of the JB language. The first, a minor problem, is that a JB integer-list will not define a correct program when a loop is created among the Auxiliary Statements. In practice, with a moderate number of AS's this is a rare occurence. Moreover, it is easy to remove such programs during the expansion of the body of the program. (In any case, this problem will be removed in the TB language below.)

A second, more serious problem is that while the mechanisms of the applications of the GOs are very simple in the JB language, the semantic implications of their use are quite complicated. Because of the structure of JB, semantic positioning of a integer-list element is extremely sensitive to change. As a specific example, consider a large complicated program beginning with a :BLOCK statement in the top-level Main Statement. A single, unfortunate, mutation converting this operator to a :SET would destroy any useful features of the program. Secondly, this strongly epistatic nature of JB seems incompatible with Crossover, given Crossover's useful-feature-passing nature. A useful JB substructure shifted one integer to the right will almost certainly contain none of its previously useful properties.

THE TB LANGUAGE

In an effort to alleviate these problems, we consider a modified version of JB. This language, called TB, takes advantage of the implicit tree-like nature of JB programs.

TB is fundamentally the same as JB except that the Auxiliary Statements are no longer used. Instead, when a TB statement is generated, either at its initial creation or as a result of the application of a GO (defined below), any subsidiary statements that the generated statement contains are recursively expanded at that time. The TB programs no longer have the simple list structure of JB, but instead are tree-like. Because we are simply recursively expanding the internal statements without altering the actual structure of the resulting program, the TB programs still satisfy Goal 2. Indeed, it can be seen that, because of its tree-like structure, TB does not suffer from the problem of internal loops described above. Thus, all possible program trees do indeed describe syntactically correct programs.

An example of a **TB** program is:

(0 (3 5) (1 3 (1 4 (4 5))))

This expands to the same **PL** and **JB** multiplication programs given above.

The standard **GO**s are defined in the following way:

Random Mutation could be defined to be the random altering of integers in the program tree. This would be valid but would encounter the same "catastrophic minor change" problems as did **JB**. Instead, Random Mutation is restricted to the statements near the fringe of the program tree. Specifically: 1) to leaf statements, i.e., those that contain operators that do not themselves require statements as arguments (:INC, :SET, :ZERO). And 2) to non-leaf statements (with operators :BLOCK, :LOOP) whose sub-statement arguments are themselves leaf operators. Inside a statement, mutation of a variable simply means randomly changing the integer representing that variable. Mutating an operator involves randomly changing the integer representing the operator and making any necessary changes to its arguments, keeping any of the integers as arguments that are still appropriate, and recursively expanding the subsidiary statements as necessary.

Similarly, following Smith[6], we restrict the points at which Crossover can occur. Specifically, Crossover on **TB** is defined to be the exchange of subtrees between two parent programs; this is well-defined and clearly embodies the intuitive notion of Crossover as the exchange of (possibly useful) substructures. This method is also without the problems that Crossover entails in **JB**. In a similar manner, we could define Inversion to be the exchange of one or more subtrees within a given program.

EXAMPLE

As a concrete example, an attempt was made to "evolve" concise, two-input, one-output multiplication functions from a population of randomly generated functions. As discussed by Smith[3(Chap5)] a major problem here is one of "hand-crafting" the evaluation function to give partial credit to functions that, in some sense, exhibit multiplication-*like* behavior, without actually *doing* multiplication.

After much experimentation, the following scheme for giving an evaluation score was used. For a given program body to be scored, several instantiations of the function were made, each having a different pair of input variables [IVs]. Each of these test functions was given a number of pairs of input values and the values of all of the function's variables were collected as output variables [OVs]. The resulting output values were examined and compared against the various combinations of input values and IVs. The following types of behavior were noted and each successive type given more credit: 1] OVs that had changed from their initial values. (Is there any activity in the function?) 2] Simple Functional dependence of an OV on an IV. (Is the function noticing the input?) 3] The value of an IV is a factor of the value of an OV. (Are useful loop-like structures developing?) 4] Multiplication. (Is an OV exactly the product of two IVs.)

Furthermore, rather than accept input and/or output in arbitrary variables, scores were given an extra weight if the input and/or output occurred in the specific target variables. To ensure that the functions remain reasonably short, functions beyond a certain length are penalized harshly. Finally, a limit is placed on the length of time a function is permitted to run; any function that has not halted within in this time is aborted.

186

A number of test runs were made for the system with a population size of fifty. These were compared against a set of control runs. The control runs were the same as the regular runs except that there was no partial credit given; all members of the population were given a low, nominal score until they actually started multiplying correctly. All runs were halted at the thirtieth generation. The system produced the desired multiplcation functions 72% more often than the control sample.

FUTURE WORK

Finally, a number of questions remain concerning the present system and its various extensions:

Extensions of the Present System: Generation of other types of simple arithmetic operations seem to be the next step in this direction. Given the looping nature of the underlying **PL** language it seems obvious that the system should be well suited for also generating addition functions. However, it is less clear that it would do equally well attempting to generate, e.g., subtraction or division functions, to say nothing of more complicated mathematical functions. Indeed, the results of the control case above show that it is difficult not to produce multiplication in this language; generation of other types of functions would prove an interesting result. On the other hand, are there other, comparably simple, languages that are better suited to other types of functions?

Concerning Extensions of the Language: A useful feature of the original **JB** language is its suitability for the mechanisms of the **GOs**. Can some further modification be made to the current **TB** language to bring it back into line with a more traditional bit-string representation? Are these modifications, in fact, really desirable? Alternatively, would it be useful to modify the languages to make **GOs** less standard? For example, would it be productive to formalize the subroutine swapping nature of the present method of Crossover and define a program as a structure comprising a number of subroutines, where the application Crossover and Inversion was restricted to the swapping of entire subroutines, and Random Mutation restricted to occurring inside the body of a subroutine?

ACKNOWLEDGEMENTS

I would like to thank Dr. Dave Davis for innumerable valuable discussions and Dr. Bruce Anderson for preserving the environment that made this work possible.

REFERENCES

1. Holland, John H., **Adaptation in Natural and Artificial Systems**, Univerity of Michigan Press, 1975.
2. Bethke, A., *Genetic Algorithms as Function Optimizers*, Ph.D. Thesis, University of Michigan, 1980.
3. Smith, S.F., *A Learning System Based on Genetic Adaptive Algorithms*, Ph.D. Thesis, Univ. of Pittsburghm, December, 1980.
4. Holland, J.H. and J. Reitman, *Cognitive Systems Based on Adaptive Algorithms*, in **Pattern Directed Inference Systems**, Waterman and Hayes-Roth, Ed. Academic Press, 1978.
5. Brainerd, W.S. and Landweber L.H., **Theory of Computation**, Wiley-Interscience, 1974.
6. Smith, S.F., *Flexible Learning of Problem Solving Hueristics through Adaptive Search*, Proc. IJCAI-83, 1983.

ADAPTIVE "CORTICAL" PATTERN RECOGNITION

by

Stewart W. Wilson

Rowland Institute for Science, Cambridge MA 02142

ABSTRACT

It is shown that a certain model of the primate retino-cortical mapping "sees" all centered objects with the same "object-resolution", or number of distinct signals, independent of apparent size. In an artificial system, this property would permit recognition of patterns using templates in a cortex-like space. It is suggested that with an adaptive production system such as Holland's classifier system, the recognition process could be made self-organizing.

INTRODUCTION

Templates are generally felt to have limited usefulness for visual pattern recognition. Though they provide a simple and compact description of shape, templates cannot directly deal with objects that, as is common, vary in real or apparent (*i.e.*, imaged) size. However, the human visual system, in the step from retina to cortex, appears to perform an automatic size-normalizing transformation of the retinal image. This suggests that pattern recognition using templates may occur in the cortex, and that artificial systems having a similar transformation should be investigated. Properties of the retino-cortical mapping which are relevant to pattern recognition are discussed in the first half of this paper. In the second half, we outline how an adaptive production system having template-like conditions might recognize patterns that had been transformed to a "cortical" space.

THE RETINO-CORTICAL MAPPING

Recent papers in image processing and display, and in theoretical neurophysiology, have drawn attention to a nonlinear visual field representation which resembles the primate retino-cortical system. Weiman and Chaikin [1] propose a computer architecture for picture processing based on the complex logarithmic mapping, the formal properties of which they analyze extensively. They and also Schwartz [2]

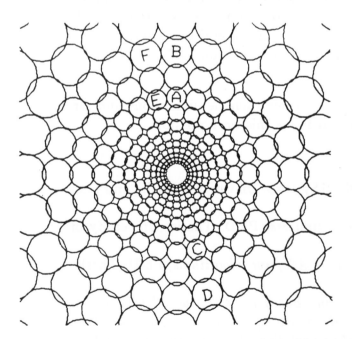

Figure 1. "Retina" consisting of "data fields" each connected to an "MSU" in the "cortex" of Fig. 2.

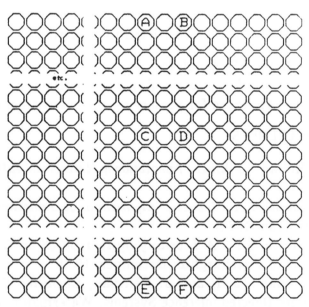

Figure 2. Each MSU receives signals from a data field in Fig. 1. Letters indicate connection pattern.

present physiological and perceptual evidence that the mapping from retina to (striate) cortex embodies the same function. Wilson [3] discusses the mapping in the light of additional evidence and examines its potential for pattern recognition. Early related ideas in the pattern recognition literature can be found in Harmon's [4] recognizer and in certain patents [5].

A hypothetical structure (adapted from [3]) schematizing important aspects of the retino-cortical (R-C) mapping is shown in Figures 1 and 2. The "retina" of Figure 1 consists of "data fields" whose size and spacing increase linearly with distance from the center of vision. The "cortex" of Figure 2 is a matrix of identical "message-sending units" (MSUs) each of which receives signals from its own retinal data field, processes the signals, and generates a relatively simple output message that summarizes the overall pattern of light stimulus falling on the data field. The MSU's output message is drawn from a small vocabulary, *i.e.*, the MSU's input-output transform is highly information-reducing and probably spatially nonlinear.

Further, all MSUs are regarded as computing the same transform, except for scale. That is, if two data fields differ in size by a factor of d, and their luminance inputs have the same spatial pattern except for a scale factor of d, then the output messages from the associated MSUs will be identical. (Physiologically, the cortical *hypercolumns* [6] are hypothesized in [3] to have the above MSU properties.)

The pattern of connections from retina to cortex is as suggested by the letters in Figures 1 and 2. Data fields along a ray from center to periphery map into a row of MSUs, and simultaneously, each ring of data fields maps into a column of MSUs. The leftmost column corresponds to the innermost ring, the 12 o'clock ray maps into the top row, and so forth.

It is convenient to describe position in retinal space by the complex number $z = re^{i\phi}$, where r and ϕ are polar coordinates. We can denote cortical position by $w = u + iv$, where u is the column index increasing from left to right and v is the row index increasing downwards. For the mapping to have complex logarithmic form, it must be true that the position w of the MSU whose data field is at z satisfies $w = \log z$ or, equivalently, $u = \log r$ and $v = \phi$.

That the equations do hold can be seen from Figure 1. The distance Δr from one data field center to the next is proportional to r itself, which implies that u is logarithmic in r. Similarly, the fact that all rings have equal numbers of data fields directly implies that v is linear in polar angle. Thus (with appropriate units) we have $w = \log z$. (The singularity at $z = 0$ can be handled by changing the

function within some small radius of the origin. For present purposes we are interested in the mapping's logarithmic property and will ignore this necessary "fix").

Figures 3-5 (at end of article) review three salient properties of the R-C mapping that have been noted by previous authors. The photos on the left in each figure are "retinal" (TV camera) images. On the right are crude "cortical" images obtained by the expedient of sampling the retinal data field centers. The mapping used has 64 MSUs per ring and per ray.

Figure 3 shows a clown seen at two distances differing by a factor of three. The cortical images, though "distorted", are of constant size and shape. Also shown is the result of rotating the clown through 45 degrees; again, cortical size and shape remain the same. The pictures show how retinal scale change and rotation only alter the *position* of the cortical image. Figure 4 illustrates these effects for a texture. The cortical images are again the same except for a shift. The mapping thus brings about a kind of size and rotation invariance which one would expect to be useful for pattern recognition.

Figure 5, in contrast, shows that the mapping lacks translation invariance. The same clown is seen at a constant distance but in three different positions with respect to the center of vision. Translation non-invariance would appear to be a distinct disadvantage for pattern recognition.

As the clown recedes from the center in Figure 5, its cortical image gets smaller and less defined. The effect illustrates how in a sense the mapping optimizes processing resources through a resolving power which is highest at the center and decreases toward the periphery. This variation is sometimes cited as a useful property of the eye, and was discussed in connection with an artificial retina-like structure by Sandini and Tagliasco [7].

OBJECT-RESOLUTION

The pattern recognition potential of the mapping's size-normalizing property is best seen by defining a somewhat unusual notion of resolution. Recall first that the resolving power ρ of a sensor is the number of distinct signals per unit visual angle; in the case of a linear sensor (such as a TV camera), ρ is a constant. Suppose we ask of a system: when its sensor images a centered object of half-angle A, how many distinct signals, corresponding to the object, will the sensor produce? Let us name this quantity the system's *object-resolution*, R_o. Then, in the case of a linear system, it is clear that R_o will be proportional to $\rho^2 A^2$. That is, R_o will depend on the distance or "apparent size" of the object, or on the relationship between perceiver and object.

The resulting amount of information may be insufficient for recognition, it may be just right, or it may overload and therefore confuse the recognition process. This uncertainty leads to the scale or "grain" problem noted by Marr [8] and others and to Marr and Hildreth's [9] proposed solution of computations at several resolutions which are later to be combined. The grain problem is also a motivation for the application of relaxation techniques [10] in pattern recognition.

Let us now ask what is the object-resolution of an R-C system. For such a system the resolving power is $\rho = c/r$, with r the distance from the center of vision. The constant c can be defined as the number of MSU outputs per unit visual angle at an eccentricity of $r = 1$. Object-resolution R_o can be found by taking a centered object of half-angle A and integrating over the object from a small inner radius ϵA ($\epsilon \ll 1$) out to A. We have

$$R_o = \int_{\epsilon A}^{A} \frac{c^2}{r^2} 2\pi r dr = 2\pi c^2 ln \frac{A}{\epsilon A} = 2\pi C^2 ln \frac{1}{\epsilon} \quad ,$$

independent of A.

Thus the mapping's object-resolution or spatial quantization of the seen object is independent of the object's apparent size or distance, and independent of its actual size as well. It depends only on c (and ϵ). Given a fixed value of c, the system may be said to see every centered object, regardless of size, equally well, independent of the perceiver-object relationship. (Strictly speaking, the above integral includes only a fraction $1 - \epsilon^2$ of the object, the "outer" fraction. But if ϵ is very small the omitted fraction ϵ^2 will contain an insignificant portion of the object's pattern.)

The object-resolution of the R-C mapping can be thought of in terms of the number of data fields per retinal ring. By mentally superimposing and then expanding and contracting a centered object on Figure 1, one can see that it is examined in an equivalent way at any scale. In fact, it is convenient to use the number of fields per ring as a measure of R_o.

The R-C mapping's *constant* object-resolution is the significant difference between it and a linear system. In the remainder of the paper we will develop implications of this difference. First, why in an important sense the "grain" problem disappears. Second, why Gestalt-like templates are, cortically, suitable for pattern recognition. Third, in outline, how the cortical approach with templates allows a separate adaptive theory due to Holland [11] to be applied to pattern recognition—and in the process may solve the mapping's apparent problem of translation non-invariance.

THE "GRAIN" PROBLEM

Basically, a "grain" problem exists if there is no *a priori* way to tell whether the size of the elements with which the perceiver is looking is the same as that of the optimally informative element of the object or scene. In the linear case, we found that the information about an object may be insufficient, just right, or overloading depending on (1) the perceiver-object relationship and of course on (2) the amount of detail in the object itself.

In the R-C mapping case, the information is *constant*, dependent only on the perceiver. Thus (1) above—uncertainty due to the perceiver-object relationship—disappears. But the information may still, it seems, be insufficient, just right, or overloading—depending on object detail.

We can develop a criterion for the latter as follows. Let an object's "object frequency spectrum" be the two-dimensional Fourier spectrum of a geometrically similar object of unit size, and let f_o be the highest significant (for discrimination) frequency in such a spectrum. Then, roughly, we may say that a mapping with resolution R_o (in units of fields per ring) provides sufficient information about an object if $R_o \geq f_o$.

But this bound is not ultimately limiting. It only says whether information from *one fixation* is sufficient for recognition. Peculiarly, by the mapping's constancy of information, any fixated local *part* of an object is seen in as much detail as is the whole object. Thus, if $R_o < f_o$, the system can always *gather* enough information by scanning, *i.e.*, by moving the center of fixation to any part not seen clearly. R_o is therefore always sufficient, though several fixations may be required.

Can there be too much resolution? Only if objects turn out to be simpler than expected. But often this can be known in advance. In contrast, in the linear case, superfluous resolution will always occur whenever object images become large.

TEMPLATES

In any digital computer implementation, a template for pattern matching consists of a finite (usually rectangular) array of cells in each of which the relative brightness to be matched is specified. The array has a fixed resolution since the number of cells is fixed.

One major traditional problem with templates is a variation of the "grain" problem: Unless the template's resolution is the same as the system's object-resolution, there is virtually no chance of getting a correct match. The R-C mapping offers a solution since the system's object-resolution is fixed, and the resolution of all stored templates can be made exactly commensurate. For instance, the system can acquire its templates by copying its own cortical MSU output images of identified objects. The same objects when later presented in other sizes will be "seen" in the same way.

Templates have other problems, e.g., orientation and brightness variations may lead to mismatch. These will be taken up later. Our analysis suggests, however, that templates may yet have an important role to play in general pattern recognition, provided the matching occurs in a cortex-like space.

OUTLINE OF AN ADAPTIVE CORTICAL PATTERN RECOGNITION SYSTEM

This section will outline a system concept combining the R-C mapping, a production system based on cortical templates, and the theory of adaptation due to Holland.

A visual world mapped as in Figures 1 and 2 suggests a natural polarity between center and periphery. The same centered object, as it grows bigger, expands toward the periphery, and its cortical image, as noted, shifts as a unit from the left side of the "cortex" toward the right side. The implication is strong that processing, in the cortex, should consist of a column-by-column scan [12] from left to right. The pattern of an object, whatever its degree of shift from the left, will be encountered "sooner or later" and thus be available for matching against templates.

Further reflection suggests that rather than working with two-dimensional templates, it might be simpler to use one-dimensional column templates—the identification of a pattern consisting of successive matching of the appropriate column templates. Storage would be saved because a given column template would often be a contributor in more than one two-dimensional match.

An appropriate structure for performing the correlation of successively matching column templates is a form of production system in which (1) the condition of each production includes a column template pattern and one or more internal message patterns, and (2) the action is an internal message to be placed on the common message list. (These internal messages are distinct from the MSU output messages. To avoid confusion, the internal messages will be called i-messages.)

In addition, a separate set of "effector" productions, whose conditions consisted only of i-message patterns, would monitor the i-message list. When an appropriate i-message appeared on the list, the effector would fire. Its "action" would be (1) an external action such as moving the center of vision, or (2) an "internal" action also modifying the system's

frame of reference but not directly observable from the outside (more on this later), or (3) a signal to the outside world denoting a pattern name.

Many details need to be filled in to make this an operating system. However, enough has been given to suggest a process in which starting at the left end of the cortex, columns would be scanned and productions would fire in dependent sequence (the dependency based on i-messages as well as the column information being matched), resulting ultimately in an effector firing whose signal named the object in view.

Production systems have not usually been considered in connection with pattern recognition because production conditions typically deal with "normalized" or logical variables and, given the grain problem, patterns in linear vision are anything but normalized. In cortical space, however, patterns *are* normalized so that there the power of productions can potentially be exploited.

But we can go farther. One part of the adaptive theory due to Holland is concerned with "cognitive systems" based on sets of productions called "classifiers". The form of a classifier is, most generally, a string whose condition part consists of a fixed length "environmental detector pattern" together with one or more i-message patterns, and whose action part is an output i-message or effector action. The important point for us is that the "environmental detector pattern" has exactly the form of the column templates we have been considering, so that classifier systems and the adaptive theory may be directly applicable to "cortical" pattern recognition. It has been demonstrated [13-16] that given an appropriate external reward regime a classifier system can evolve a set of classifiers that is adapted to, or "fit", in its environment. This means in particular that the conditions of the classifiers recognize what matters, and the i-messages and actions are appropriate. Much further research must be done, but by combining classifiers with R-C vision, a new path would appear to be open to the objective of a self-organizing visual pattern recognition system.

If the adaptive properties of the Holland system be assumed, we can suggest how the production structure given earlier might deal with non-centered objects. They look different from their centered forms: this is the mapping's translation non-invariance. The problem would be solved if classifiers existed which would react to the off-center form and lead to an effector which would move the center of vision so as to center the object (at which point "standard" classifiers could recognize it).

At first sight, the evolution of this kind of sequence seems implausible: you would need classifiers for every object in every peripheral position. However, the mapping helps by *reducing the detail* seen in an object as it recedes toward the periphery; in the limit, every object becomes just a "blob". This suggests that only a relatively small number of distinct classifiers would be needed to "acquire" any object for standard (centered) inspection.

There remains the problem, not of the isolated object, but of the more-or-less centered one—such as a face—which is still not centered quite well enough to fire its standard classifiers. How can an appropriate centering movement come about? For this question, and related ones, we need to consider the "internal effectors" mentioned earlier.

Three are important in the present discussion: Object-Resolution (OBRES), Azimuth (AZIM), and Brightness Gain (BGAIN). OBRES is an effector (or set of them) which, given appropriate i-messages, will *alter* the system's object-resolution (in effect changing the number of data fields per ring in Figure 1). This permits seeing an object (regardless, of course, of its apparent size) in detail, or more coarsely, depending on the i-message list circumstances. The evolution of OBRES effectors appropriate to different circumstances would occur through the adaptive mechanisms.

If we now recall the problem of the slightly off-center face, it seems plausible that, given some reduced level of object-resolution, most different faces with that degree of decentering could be matched by a relatively small (and thus practical) set of classifiers. These would lead to a movement command bringing the face to the center, where it would be recognized in detail (after, perhaps, restoration by OBRES of a higher R_o).

The AZIM internal effectors set the direction the system regards as "up". In cortical space, this amounts to shifting the input column vector along its length by a definite amount before matching classifier template patterns against it. The purpose of AZIM is, of course, to allow a given set of classifiers to be effective for recognition even if the object is not in standard orientation. But how will the right azimuth be set in such a case? We again have recourse to the evolution of relatively coarse classifiers which, given reduced object-resolution through OBRES, will recognize the presence of a nonspecific ("oblong", say) object at a certain orientation. These would lead to the right AZIM acting, and specific recognition could then occur.

Finally, BGAIN is a set of internal effectors to deal with the persistent problem of setting the right brightness level for template matching. The intent is that the appropriate gain will be determined (via the i-message list) by what is seen, and that the evolution of an appropriate set of BGAIN effectors will again be under adaptive control in the Holland

sense.

The various internal effectors, and the external one resulting in movement, are concerned with the system's "point of view" on its visual input, that is, with systematic transformations which will allow the system's form detector set—the classifiers—to function efficiently.

SUMMARY

We began this paper with the retino-cortical mapping and showed how it "saw" centered objects with a resolution independent of the object's size. Constant object-resolution led to a renewed prospect for template matching in general pattern recognition. Fixed size templates permitted the power of production systems to be brought to bear. Finally, the applicability of Holland's adaptive theory to production systems allowed us to suggest that a recognition system based on the mapping might be made self-organizing, in the process overcoming the mapping's "problem" of translation non-invariance.

REFERENCES

[1] Weiman, C.F.R. & Chaikin, G. Logarithmic spiral grids for image processing and display. *Computer Graphics and Image Processing*, 11, 197-226. 1979.

[2] Schwartz, E.L. Spatial mapping in the primate sensory projection. *Biological Cybernetics*, 25, 181-194, 1977.

[3] Wilson, S.W. On the retino-cortical mapping. *Int. J. Man-Machine Studies*, 18, 361-389, 1983.

[4] Harmon, L.D. Line-drawing pattern recognizer. *Electronics*, 39-43, Sept. 2, 1960.

[5] Singer, J.R. Electronic recognition.
U.S. 3,255,437, Jan. 7, 1966.
Burckhardt, C.B., *et al.* Pattern recognition apparatus utilizing complex spatial filtering.
U.S. 3,435,244, March 25, 1969.
McLaughlin, J.A., *et al.* Pattern recognition apparatus and methods invariant to translation, scale change, and rotation.
U.S. 3,614,736, October 19, 1971.

[6] Hubel D.H. & Wiesel, T.N. Uniformity of monkey striate cortex: a parallel relationship between field size, scatter, and magnification factor. *J. Comp. Neurology*, 158(3), 295-305. 1974.

[7] Sandini, G. & Tagliasco, V. An anthropomorphic retina-like structure for scene analysis. *Computer Graphics and Image Processing*, 14, 365-372, 1980.

[8] Marr, D. Early processing of visual information.

Philosophical Transactions of the Royal Society of London B, **275**, 483-524, 1976.

[9] Marr, D., & Hildreth, E. Theory of edge detection. *Proc. Royal Society of London B*, **207**, 187-219, 1980.

[10] Davis, L.S. & Rosenfeld, A. Cooperating processes for low-level vision: a survey. *Artificial Intelligence*, **17**, 245-263, 1981.

[11] Holland, J.H. *Adaptation in Natural and Artificial Systems*. Ann Arbor: U. of Michigan Press, 1975.

[12] Evidence and a model for scanning in humans is presented in Wilson, S.W., Strobe imagery: a scanning model. Submitted for publication.

[13] Holland, J.H., & Reitman, J.S. Cognitive system based on adaptive algorithms. In *Pattern-Directed Inference Systems*, Waterman, D.A. & Hayes-Roth, F. (eds.). New York: Academic Press, 1978.

[14] Booker. L. *Intelligent behavior as an adaptation to the task environment*. Ph.D. Dissertation (Computer and Communication Sciences). The University of Michigan, 1982.

[15] Goldberg, D.E. *Computer-aided gas pipeline operation using genetic algorithms and rule learning*. Ph.D. Dissertation (Civil Engineering), The University of Michigan, 1983.

[16] Wilson, S.W. Knowledge growth in an artificial animal. These *Proceedings*.

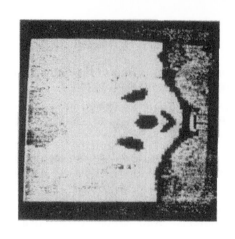
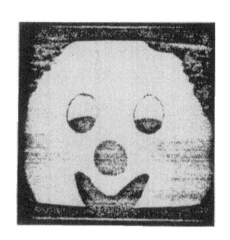
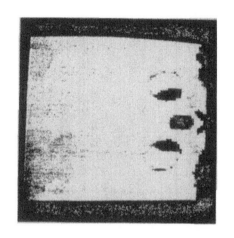

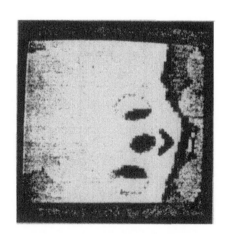

<u>Fig. 3</u>

194

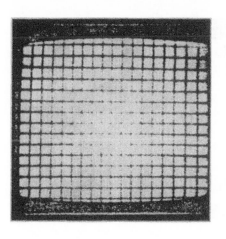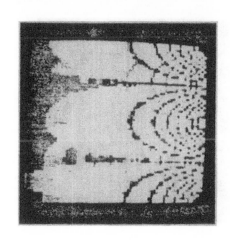

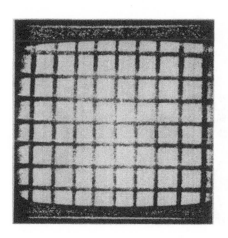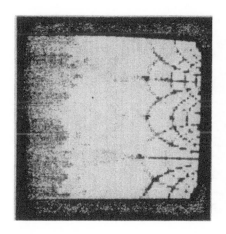

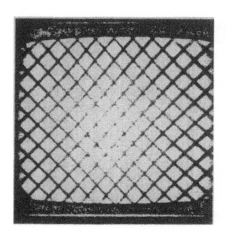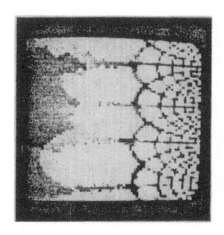

<u>Fig. 4</u>

195

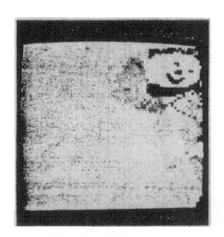

Fig. 5

196

MACHINE LEARNING OF VISUAL RECOGNITION USING GENETIC ALGORITHMS

Arnold C. Englander
Itran Corporation, Manchester, N.H.

ABSTRACT

This paper briefly describes preliminary work with an application of genetic algorithms. Genetic algorithms are used as the mechanism by which a vision recognition system learns to classify distorted examples of different but similar classes of image patterns. The system develops increasingly effective collections of class specific feature detectors producing increasingly unambiguous, hence reliable, recognition performance. Algorithms and early simulation results are described.

Genetic algorithms are applied to a special case of a difficult optimization problem which is emerging in several forms in computational vision research. The general optimization problem has a performance measure that is easily formulated as an algorithm involving the composition of both functionals and logical operations. However, the performance measure is not itself a smooth, much less convex, functional. This precludes the application of most conventional optimization techniques.

I. INTRODUCTION

A variety of techniques for the machine recognition of objects in images exist in the literature and in demonstrated machine vision technology [1,2,3]. There is an image recognition problem which is difficult for all of these techniques but which arises in practical applications. The problem combines two troublesome characteristics. First, pattern classes have prototypes which correlate highly with the prototypes of different pattern classes. Second, the pattern examples (to be classified) are randomly distorted and occluded. Practical cases of this prOblem arise in reading characters stamped in certain industrial materials such as rubber and cast metal. Other examples are found in robot vision "bin-picking" applications involving certain assortments of parts. This paper describes the use of genetic algorithms as the basis of a machine vision system which improves its own performance with such recognition problems by learning from labeled examples.[1]

II. THE OPTIMIZATION PROBLEM

Experience in applying conventional recognition techniques to difficult industrial vision problems has led to this view: Robust recognition performance relies on the identification and use

---

[1] For a general and thorough introduction to genetic algorithms, including general analytical results, see the pioneering book by Holland [4].

of a large set of local image features having two properties. First, important local features are those which, either alone or in small groups, disambiguate the recognition process by being necessary and/or sufficient ("essential") evidence for classification. Second, such features and groups of features must be likely survivors of the distortion and occlusion operations under which image pattern examples are generated from class prototypes.

Obviously essential features are application dependent. They depend on the class prototypes and on the distorting and occluding processes. The problem's strong dependence on application particulars leads to the requirement that the recognition system improve its own performance by associative learning from labeled examples.

It is desirable to identify many small features which are essential when detected alone or in a variety of groupings. This way the features which contribute to the recognition process are likely to survive the random distortions and occlusions. The detections of essential features should be not only graded and combined in weighted sums but combined in ways which allow pieces of evidence to "veto" the significance of other pieces of evidence. Intuitively, the behavior of algorithms based on such ideas will be complicated by implicit non-

linear, "competitive" and "cooperative" interactions between the evidence derived from the detections of essential features.

II. USE OF GENETIC ALGORITHMS

Applying these views to machine learning of visual recognition leads to an optimization problem over a space of populations of 2-D detector arrays where each array is a composite of templates for the detection of essential image features. The overall population of detector arrays is divided into class specific sub-populations each of which is optimized to respond maximally to examples of a particular image pattern class. The recognition algorithm classifies unidentified images by assigning them to the detector array sub-population producing the highest sum of individual recognition responses. The recognition response of an individual detector is the product of a match between the detector and the input image, and a term called "strength". The strength of a detector array is indicative of the detector array's past performance in disambiguating recognition decisions.

Optimization of a sub-population of class specific detector arrays means finding detectors which strongly match input image examples of the specified class, but which only weakly match input image examples of other classes. This

198

is difficult because the different image pattern classes have prototypes which are alike in the sense of being highly cross-correlated. This optimization problem reflects the desired strategy and intuitively seems simple. However, it is not easy to solve. The problem's performance measure on individual detector arryas is composed of functionals and logical operations. It is not itself a smooth, much less convex, functional. Such optimization problems are unsolvable by most conventional methods. Because genetic algorithms impose unusually few constraints on the formulation of optimization problems they are applicable to this problem.[2]

The match between detectors and input images involves a "matchscore" which is common to most genetic algorithms. The strength of detectors develops iteratively. During the associative learning phase of the system, the strength of each detector is increased each time the detector's response is above the average response of all detectors and the class origin of the input image and the class specificity assignment of the detector are the same. The strength of a detector is decreased each time it produces an above average response to an input image

originating from a class other than the class to which the detector's sub-population is being optimized to recognize.

Here, an image pattern is a 2-D array of binary valued picture elements, or "pixels". (This corresponds to a 2-D map of the zero crossings in a digital image processed by convolution with a difference of gaussians (DOG) operator for the detection of edges. The resulting zero crossings are useful in portraying the boundaries of objects in the scene.) The image patterns are randomly distorted and occluded examples of prototypes from one of several distinct, but similar, image pattern classes.

A detector array is a 2-D array of pixels of the same size as the image patterns. Here each pixel takes one of three symbols, {0,1,#} where {0,1} indicate values taken by pixels in image patterns and # indicates the "don't care" condition in the usual genetic algorithm matchscore. A standard matchscore is used in mating image patterns to detectors arrays by simply "unwinding" the image patterns and detector as taxa type character strings (over {0,1} for image patterns and over {0,1,#} for detectors).

Genetic algorithms optimize the class specific sub-populations of detector arrays, indirectly,

2   Other cases of such optimation problems are emerging in computational vision research [5]. One case involves the goal of combining the information of various visual processes (stereopsis, motion, and "shape from-shading" for example) into a single interpretation (of 3-D or "2-1/2-D" for example), which is optimal under a performance measure which combines functionals and logic. Genetic algorithms may be applicable to such problems as well.

by operating on the individual detector arrays in each separate, class-specific sub-population. Restricting "mating" and "replacement" operations to taxa within the same sub-population, two "parents" are selected (in each sub-population, at the completion of each recognition trial involving labeled examples, hence changes in strengths). The "parent" taxa are selected according to the detectors returning the two highest recognition responses (the product of the match with the current input image example and the detector strength) or with probabilities proportional to the recognition responses. The two "parents" generate two "offspring" under genetic operators and the "offspring" each replace an "individual" judged to be "weak" for having one of the two lowest strengths of the taxa in the sub-population. The "offspring" enter the sub-population with strengths which are a fraction of the average strength of the two "parents" and the strengths of the "parents" are reduced to match that of their "offspring".

These selection rules reflect heuristic arguments and experimentation. "Parents" are selected as to recognition responses to ensure that they are "strong" for having contributed to disambiguation in the past, and that they are well matched to the current input example. "Weak" individuals are "un-selected" by low "strength" alone, rather than

by the current match-"strength" product, to avoid losing detector arrays which tend to be useful but match poorly with the current input example (which is randomly distorted and occluded).

Early simulations involved standard operators of genetic algorithms: "cloning", "cross-over", "inversion", and "muta-tion", chosen according to pro-babilities which are fixed for each experiment. As is commonly believed, it is most useful to assign "crossover" the highest usage probability. Experiments were also performed using Wilson's "imprinting" and "ternary intersection" opera-tors, with low usage probabili-ties. Wilson's operators seem relevant and useful to this problem [6].

III. EARLY SIMULATION RESULTS

Early simulation results are promising in that self-optimi-zation by genetic algorithms is obvious. The recognition system, operating in training mode, clearly improves its cumulative average of correct recognitions from very low initial percentages to moderately high percentages over a few hundred trials. In simulations involving 4 pattern classes of 2 prototypes each, 4 sub-populations of detector arrays having 32 detector arrays each, and image and detector arrays of 32 by 32 pixels, the system averaged correct recognitions 25% of the time for the first 100 or so trials, rising exponentially to 78% correct recognitions after

1000 trials. In such simulations the detectors were initialized with pixels containing 0,1,#, with equal probability and Wilson's genetic operators were used randomly with small probabilities. In some simulations the system improved its recognition performance over correlation based pattern recognition techniques in a few thousand training iterations.

As expected, over time, the system evolves strong detector arrays which partly resemble the prototypes of the pattern classes to which the detectors are assigned. But the resemblance is never complete because detectors must match features present in examples of their assigned pattern class but ignore features which are also characteristic of other classes. The evolution of such detectors is apparent in the simulations.

IV. CONCLUSION

Preliminary work with an application of genetic algorithms has been described. Genetic algorithms are the mechanism by which a vision recognition system learns to classify distorted examples of different but similar classes of image patterns. This work addresses an unconventional optimization problem which arises naturally from an intuitive model of visual learning. Early simulation results indicate that the proposed model can lead to the design of an effective machine vision system.

REFERENCES

1. R. Duda, P. Hart: <u>Pattern Classification and Scene Analysis,</u> Wiley, New York, 1973.

2. E. Hall: <u>Computer Image Processing and Recognition</u>, Academic, New York, 1979.

3. J. Tou and R. Gonzalez: <u>Pattern Recognition Principles</u>, Addison-Wesley, Reading, MA, 1974.

4. J. Holland: <u>Adaption in Natural and Artificial Systems</u>, University of Michigan, Ann Arbor, 1975.

5. D. Terzopoulos: "Multilevel Reconstruction of Visual Surfaces: Variational Principles and Finite-Element Representations", in <u>Multiresolution Image Processing and Analysis</u>, ed. A. Rosenfeld, Springer, New York, 1984 (see page 283).

6. S. Wilson: "Knowledge Growth in an Artifical Animal", in Proc. Fourth Yale Workshop on Applications of Adaptive Systems Theory, New Haven, Conn., 1985.

Bin Packing With Adaptive Search
Derek Smith
Texas Instruments

1.0 INTRODUCTION

We have looked at the problem of bin packing arbitrarily dimensioned rectangular boxes into a single orthogonal bin. Figure 1 shows a good bin packing, the sort we are aiming for. Figure 2 shows a poor bin packing.

The problem is NP-hard in the strong sense, so there is little hope of finding a polynomial time optimisation algorithm for it (1). Reasonable approximation algorithms exist which can be guaranteed to be within 22% of optimal (1).

Our approach has been to use a wrinkle on genetic algorithms (3), developed in the Texas Instruments Computer Science Laboratory (2).

2.0 ADAPTIVE SEARCH

The epistatic domain of bin packing has traditionally not been amenable to adaptive search techniques. This is because it is difficult to represent a bin packing on which we can do crossover and mutation and retain either a reasonable packing or a legal packing.

Consider a flip mutation (rotate through 90 degrees) of box 18 in figure 1. The flip will either cause a illegal bin packing due to boxes overlapping each other, or if we fracture the packing by moving the neighbouring boxes away to make the flip legal, will produce a poor bin packing.

Our solution is to represent the bin packing as a list of the boxes plus an algorithm for decoding the list into a bin packing. The list is readily mutatable (flipping boxes), and is amenable to a modified form of crossover. The decoding algorithm takes any list of boxes and forms a legal packing. Hence we attempt to produce good bin packings using Genetic Algorithms.

2.1 The Representation

As explained above our representation is a list with an associated algorithm to apply to the list to produce a bin packing. For effective search the algorithm must produce legal packings from any operation on the list. Here we describe two such decoding algorithms.

The first algorithm we call SLIDE PACK. We take each box, in order, from the list, place it in one corner of the bin, and let it fall to the farthest corner away, as if under a gravity that only allowed it to move orthogonally. The effect is that a box will zigzag into a stable position in the opposite corner from which it was placed. Box 2 in figure 3 shows the SLIDE PACK algorithm.

SLIDE PACK is fast as there is no backtracking, and is simple to compute. Its time complexity is $O(n**2)$, where n is the number of boxes. There are n! possible orderings of our list of n boxes. If we associate a flipped state with each box, this gives us $n!2**n$ members in the set of all encoded representations. Although we can contrive packings that SLIDE PACK can never do, we believe that in general we can reach all of the search space by operating on the list of boxes.

The second algorithm we call SKYLINE PACK. For each box in the list, in order, we try the box in all stable positions, and in all its orientations on the partially packed bin. A stable position is where the box is tucked into a corner, or cave formed by other previously packed boxes. The algorithm takes its name from the fact that it tours the skyline formed by the previously packed boxes to find the position it fits best. Figure 4 shows some of the places that box 2 is being considered for by the SKYLINE PACKer.

Again we have n! possible orderings of the list. However each time a we pack a box we try that box in many positions - we are covering more of the search space than in the SLIDE PACKing of a box. It is clear that we can no longer generate all possible bin packings, as a poor placement of a box will be ignored in favour of a better placement somewhere else on the skyline. A more practical question is whether we can represent all good bin packings. We believe so (again informally) but with less conviction than with the SLIDE PACK. SKYLINE PACK has time complexity $O(n**4)$.

With a randomly generated list SKYLINE PACK will tend to generate a significantly denser packing than SLIDE PACK, however, it takes longer to run. Figure 2 is a typical SLIDE PACKing of a randomly generated list, whilst figure 5 is a typical SKYLINE PACKing. SLIDE PACK can produce good packings as shown in figure 1 when we apply the adaptive search techniques. The trade off is whether to run the adaptive search with larger populations and for more generations using SLIDE PACK, or in the same amount of time use SKYLINE PACK for fewer generations. Our experiments have shown that SKYLINE PACK is more favorable, however with a better tuning of the adaptive search SLIDE PACK may produce better results.

2.2 The Genetic Operators

Our representation of a packing, as described, is the order of the boxes presented to the packing algorithm. Traditional crossover cannot operate on such a list. Consider a crossover of list (1 2 3 4 5) with (5 4 3 2 1) the crossover point being after the second element to produce (1 2 3 2 1). The list now has boxes 1 and 2 duplicated and boxes 4 and 5 missing.

Hence we use a MODIFIED CROSSOVER which takes the order of the boxes before the splice from the first list, and the order of the boxes which remain to be packed from the second list after the splice point. In the above example we would generate the list (1 2 5 4 3).

Hollands theorems (3) regarding the effectivness of crossover no longer hold. We have not yet investigated the theoretical aspect of the modified crossover. However we have experimented with its use; we have run random search versus our genetic operators, and have found the genetic operators to produce consistently better results.

One of the mutations we have experimented with is SCRAMBLE, that is randomly reordering some portion of the list. At the beginning of the adaptive search process we can concentrate on SCRAMBLing the beginning portions of the list to evolve a good basis for the packing. As the evolution proceeds we can move our area of interest father up the list.

A FLIP mutation to try different orientations of the boxes is necessary if the decoding algorithm does not try the box it is packing in all its orientations. FLIP is applied discretely to boxes in the list.

2.3 The Evaluation

Because we require our evaluation procedure to score dense packings highly, a straightforward evaluation criteria is the ratio of the area of the boxes packed to the area of the bin. This works well as an evaluation of a packing.

It is less clear how to evaluate partial packings which are required in such decoding algorithms as the SKYLINE PACKer where we need an evaluation of the packing for each position of the box along the skyline, to choose where to settle it. We have tried numerous ways to measure partial bin packings. One of the most intriguing is to take the inverse square of the separation of the box being packed to all the other boxes. This favors boxes filling in caves, especially if they fit snugly into the cave. There is some analogy here to gravitational effects, and indeed such an evaluation allows us to pack in space (as opposed to in a containing bin) as the boxes are attracted to each other.

Graph 1 shows how the density of a partial bin packing falls as the number of boxes packed increases. This is due to the forming of more and larger caves by the later boxes. As the evolution continues we form less caves, and we can see from the graph that by generation 20 we have kept to about 85% density.

3.0 RESULTS

We have benchmarked our results against a recently developed deterministic bin packing program within our group. This program uses some heuristics and dynamic programming techniques. Our program can produce the same packing density 300 times faster. Also if a greater density is required then we can simply allow our program to run for longer, or run it again. Similarly if a less dense packing is required we run for only a short time. Graph 2 shows how the density increases as the evolution proceeds. This is a tremendous practical advantage of this approach. A practical disadvantage is that each time we run the process we will end up with a different packing.

4.0 FUTURE RESEARCH

There is work to be done in the mating of the decoding algorithm and the genetic operators. In particular, finding ways to operate a portion of a bin packing without having repercussions on the whole bin packing.

Work is also in progress in making the genetic operators robust to quantity of data, variation in dimensions of boxes, and variations in the aspect ratio of the bin.

We are also considering a process which monitors the adaptive search whilst it runs. Such a process could vary the importance of the mutations as the search proceeds. It could bring in mutations to produce diversity of the search if it were trapped at a local maxima. It could also alter the size of the population at various stages in the evolution. Currently such variations are set up at the start of a run, it would be more effective to have the process continually monitoring and adapting itself.

In order to learn how to implement the monitor process we need to study how the search space is being explored. Seeing our bin packing algorithms run by the use of graphics has been very useful in this work to date. Graph 3 shows the sort of display which we would like in order to watch the evolution, learn about the process, and write the self monitoring system we have mentioned. Numbers 1 through 4 are four of the members of the initial population. The trees sprouting from them represent the performance of their offspring. 1 was a poor initial packing and soon died away. 4 was a good packing and we can see it spawned many children in exploring its portion of the search space. Note also that 2 and 3 are allowed to evolve to maintain diversity in the search.

Graph 4 is the same concept as graph 3 in a search space that we have completley mapped out and in which we can draw the local maxima, represented by hills in the graph. We could then test new levels of operators, and different population sizes in a controlled and visible search space. Graph four shows only two dimensions of such a space, which for n boxes is n-dimensional.

205

5.0 ACKNOWLEDGEMENTS

This work is only possible because of the enthusiasm, research work, and utilities for adaptive search all provided by co-worker Lawrence Davis.

We thank the referees for their valuable comments.

6.0 REFERENCES

1. Garey and Johnson, Computers and Intractability, 1979, W. H. Freeman.

2. Lawrence Davis, Applying Adaptive Algorithms to Epistatic Domains, To appear proc. IJCAI-85.

3. John H. Holland, Adaptation in Natural and Artificial Systems, University of Michegan Press, 1975.

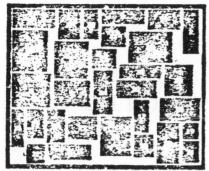

Figure 1 A good bin packing

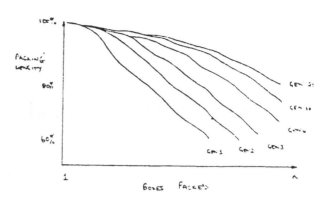

Graph 1 Density a bin packing proceeds

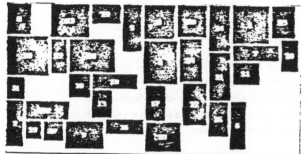

Figure 2 A typical SLIDE PACKing

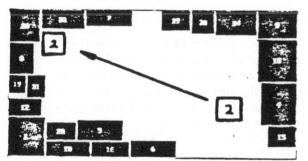

Figure 3 The SLIDE PACKer

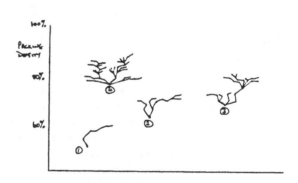

Graph 2 Density as search proceeds

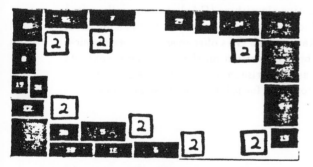

Figure 4 The SKYLINE PACKer

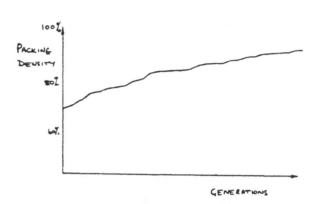

Graph 3 Tracing the evolution

Figure 5 A typical SKYLINE PACKing

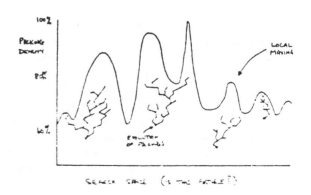

Graph 4 Exploring the search space

Directed Trees Method for Fitting a Potential Function

Craig G. Shaefer

Rowland Institute for Science, Cambridge MA 02142

Abstract

The Directed Trees Method is employed to find interpolating functions for potential energy surfaces. The mathematical algorithm underlying this fitting procedure is described along with example calculations performed using a genetic adaptive algorithm for fitting the A_k unfolding families to 1- and 2-dimensional surfaces. The properties and advantages of the usage of genetic adaptive algorithms in conjunction with the Directed Trees method are illustrated in these examples.

Section: 1. Introduction

How does one choose a mathematical model to describe a particular physical phenomenon?

To help in answering this question, we have developed a method called the Directed Trees (DT) method[1] for describing the possible structures available to a particular special type of model: the gradient dynamical systems. The gradient dynamical systems are, however, quite general and flexible and hold a ubiquitous presence in the physical sciences. In the next section we illustrate where this special type of model 'fits' into a very broad class of mathematical models. The DT method employs a relatively young branch of mathematics called differential topology: "topological" in order to form categories of solutions for gradient dynamical systems to reduce the problem to the study of a finite number of different categories, and "differential" in order to allow for quantitative calculations within these models. For the purposes of this paper, it is sufficient to say that in the numerical applications of the Directed Trees method, systems of nonlinear equations arise for which we require solutions. Although classical numerical methods could be employed for the solution of these nonlinear systems, we find that genetic adaptive algorithms (GAs) are especially suited for this purpose and have certain advantages. In order to introduce our application of GAs to the solution of nonlinear systems of equations and be able to discuss the advantages which GAs offer over the more classical numerical methods, the third section of this paper provides a brief exposition on the topological concepts inherent to the Directed Trees method and describes the equations that arise in its quantitative applications. Section 4 contains examples of the usage of genetic adaptive algorithms for solution of these systems.

Section: 2. General Mathematical Models

In this paper, we are seeking not so much a procedure for calculating the specific solution to the mathematical model of a physical system, but rather the development of a model for which we may classify its solutions into behavioral categories so that one particular solution from each category serves as a paradigm for all solutions belonging to its category. Obviously, this will greatly simplify the study of the general solution of a model. In order to do this, however, we first need to restrict the type of mathematical model to which our classification scheme is applicable. To understand where our restricted class of models fits into the general class of mathematical models, below we will describe the simplifications inherent to our restricted class. The following table contains a list of possible variables whose interrelationships we seek. These variables include items such as the spatial and time coordinates, and parameters such as the masses of particles, the refractive indices of mediums, densities, temperatures, *etc.* In addition, our model might

General Mathematical Models		
Variable	General Term	Comments
$\vec{x} \in \Re^m$	$x_i \quad (i{:}1{\to}m)$	m spatial coordinates
$t \in \Re$	t	time coordinate
$\vec{\rho} \in \Re^s$	$\rho_i \quad (i{:}1{\to}s)$	s parameters (mass, refractive index,...)
$\vec{\Theta}$	$\Theta_i(\vec{x};\vec{\rho};t) \quad (i{:}1{\to}\infty)$	**SOLUTIONS** (trajectories,...)
$\mathbf{D}_t^j \vec{\Theta}$	$\frac{d^j \Theta_i}{dt^j} \quad (i,j{:}1{\to}\infty)$	time derivatives
$\mathbf{D}_{\vec{x}}^k \vec{\Theta}$	$\frac{\partial^k \Theta_i}{\partial x_1^{\alpha_1} \cdots \partial x_m^{\alpha_m}} \quad (\alpha_1 + \cdots + \alpha_m = k)$	spatial derivatives
$\int T(\vec{\Theta})dt$	$\int T(\Theta_i;\ldots)dt \quad (T{:}\text{functional})\dagger$	time integrals of functionals
$\int K(\vec{\Theta})d\vec{x}$	$\int K(\Theta_i;\ldots)dx_j \quad (K{:}\text{functional})\dagger$	spatial integrals of functionals
$\vec{f}$	$f_\ell(\vec{x};t;\vec{\Theta};\vec{\rho};\ldots)\dagger \quad (\ell{:}1{\to}n)$	integrodifferential functionals
$\dagger$ = functionals may depend on any of the variables located above them in this table		

Table 1 Table containing possible variables, parameters, and functional dependencies for a general mathematical model of a physical system.

also depend on the derivatives with respect to the time and spatial coordinates as well as integrals whose integrands are functions of the other variables or solutions.

Suppose we have a physical system for which we have a set of n arbitrary rules that specify the interactions of the variables from Table 1. This leads to the following system of n equations, called an *Integrodifferential System*, whose solutions describe the behaviors of the physical system:

$$\vec{f} = \vec{0} \tag{1}$$

Since we have n equations, let us suppose that there are n solutions, thus we take $\vec{\Theta} = (\Theta_1, \ldots, \Theta_n)$ in what follows. Let us remark that this system forms a very general and flexible mathematical model for studying physical phenomena. It encompasses almost all mathematical models that are currently employed in the sciences. This system of integrodifferential equations is, however, much too difficult to solve in all of its generality — only in very specific cases are solutions even known. And virtually nothing is known about how these solutions vary as the parameters are changed. We must make a few simplifications in these equations before anything can be said about their general solutions. These simplifications are very typical though, for many models in the "hard" sciences have as their fundamental premises the assumptions that we describe below.

To begin, we assume that $\vec{f}$ does not explicitly depend upon $\vec{x}$, $\mathbf{D}_t^j\vec{\Theta}$ for $j > 1$, $\mathbf{D}_{\vec{x}}^k\vec{\Theta}$, $\int T(\vec{\Theta})dt$, nor $\int K(\vec{\Theta})d\vec{x}$. Then the system has the form: $\vec{f} = \vec{f}(\vec{\Theta};\vec{\rho};t;\mathbf{D}_t\vec{\Theta}) = \vec{0}$, for which more can be said concerning its solutions. Instead of studying this system, though, we continue with a further simplification concerning the dependency of $\vec{f}$ on the time derivatives, and, in particular, we consider those $\vec{f}$ of the form,

$\vec{f} = \mathbf{D}_t\bar{\Theta} - \vec{f'}(\bar{\Theta}; \vec{\rho}; t) = \vec{0}$. Note that the function $\vec{f'}$ appears to be similar to a force vector. In effect, the above system of equations describes the situation in which the rates of change of the solutions are proportional to a vector that depend upon the solutions themselves. This type of system arises in classical mechanics and is usually called a *Dynamical System*. If the further restriction that the forces do not explicitly depend on the time, then we have the following system of equations, which form an *Autonomous Dynamical System*: $\vec{f} = \mathbf{D}_t\bar{\Theta} - \vec{f'}(\bar{\Theta}; \vec{\rho}) = \vec{0}$. A few useful statements can be made about the solutions of this type of system of equations and their behaviors as the parameters $\vec{\rho}$ are varied. We, however, will again continue to make one further simplifying assumption on the form of the $\vec{f'}$. We noted above that the vector function, $\vec{f'}$, is of a form similar to the forces in kinematics and electrodynamics. If, in fact, $\vec{f'}$ is a true force, then it can be taken to be the negative gradient of some scalar potential ϕ: $\vec{f'} = -\mathbf{D}_{\bar{\Theta}}\phi(\bar{\Theta}; \vec{\rho})$. Then we have the system

$$\vec{f} = \mathbf{D}_t\bar{\Theta} + \mathbf{D}_{\bar{\Theta}}\phi(\bar{\Theta}; \vec{\rho}) = \vec{0} \qquad , \tag{2}$$

which is termed a *Gradient Dynamical System*. Many very powerful statements can be made about the $\bar{\Theta}$ and their behaviors as functions of $\vec{\rho}$ for this system. Oftentimes, we are concerned with the "stationary" solutions of (2), *i.e.*, solutions which are time-independent. These stationary solutions require the forces to vanish, in other words, we require $\mathbf{D}_{\bar{\Theta}}\phi(\bar{\Theta}; \vec{\rho}) = \vec{0}$. This equation determines what are called the *equilibria*, $\bar{\Theta}_c$, of the gradient system. The most powerful and general statements can be made about equilibria and how they depend upon their parameters.

The solutions, $\bar{\Theta}$, of the above systems are merely generalized coordinates for the physical systems, and thus, following the standard nomenclature, we replace $\bar{\Theta}$ by $\bar{x}$. For example, these solutions, $\bar{x}$, might be the positions of equilibria as functions of time, the Fourier coefficients of a time series, or even laboratory measurements.

We have thus shaved the general mathematical model, (1), of a physical process to the specific case of examining the behaviors of scalar potential functions, $\phi(\bar{x}; \vec{\rho})$. It is for these special cases that differential topology yields the most useful results.

In the next section we examine the primary results of singularity theory which allows any arbitrary potential to be classified into a finite number of different category types. It is this classification that greatly simplifies the study of gradient dynamical systems.[2] Since we are interested in the particular potential functions stemming from the solution of the Schroedinger equation, under the Born-Oppenheimer approximation, for a chemical reaction, we apply the classification scheme specifically to potential energy surfaces (PESs). Keep in mind that the same classifications and calculations are applicable, however, to any gradient dynamical system. The classification scheme that we have developed for PESs, as we have mentioned, is called the Directed Trees method and contains both a qualitative diagramatic procedure for implementing the classification as well as a quantitative computational procedure for calculation of specific behaviors and characteristics of the model.

Section: 3. The Topology of Potentials

Why should we concern ourselves with an alternate classification scheme based upon differential topology for potential energy surfaces? The reason for doing so is that this new Directed Trees classification has two special properties: structural stability and genericity.[3] The concept of structural stability plays an important role in the mathematical theory of singularities. There are several reasons for this importance. First of all, usually the problem of classifying objects is extremely difficult; it becomes much simpler if the objects one is classifying are stable. Secondly, in many cases, the class of all stable objects forms what is loosely called a generic set. This means that the set of all stable objects is both open and dense, in the mathematical sense, in the set of all objects. In other words, almost every object is a stable object and every object is "near" a stable object. Thus every object can be represented arbitrarily closely by a combination of stable objects. For instance, the Implicit Function Theorem of calculus and Sard's Theorem of differential topology imply that almost all points are regular points (points whose gradients are nonzero) for stable functions and thus are not critical points. Stated differently, regular points are *generic*, i.e., they form an open and dense subset of the set of all points for stable functions. (Stable functions are functions which can be perturbed and still maintain their same topological properties.) Even though almost all points of a function are regular points, nondegenerate isolated critical points do occur and have a generic property: they are not removed by perturbations. The importance of nondegenerate critical points extends beyond their mere existence, for they "organize" the overall shape of the function. This can be seen in the following one-dimensional example. Consider a smooth function of a single dimension, $f(x)$, which has three critical points between $x = 0$ and $x = 1$. If the curvature at the critical point with the smallest x coordinate in this interval is negative, then the curvatures for the middle and highest critical points must be positive and negative, respectively, for no other combination can lead to a smooth function connecting these three critical points. In addition, the functional values at the smallest critical point and the largest critical point must be greater than the value of the function at the middle critical point. A simple graph of f satisfying the above conditions will show that if these statements were not in fact true, then additional critical points would be required between these three critical points. Just as nondegenerate critical points organize the shape of a one-dimensional function, degenerate critical points "organize" families of functions having specific arrangements of nondegenerate critical points. These degenerate critical points are nongeneric in the sense that small perturbations either split the degenerate critical points into nondegenerate points or annihilate the degenerate point completely leaving behind only regular points. It might therefore seem that we should not concern ourselves with degenerate critical points since they are mathematically "rare" occurrences on a surface and can be removed by small perturbations. The manner in which degenerate points "organize" functions into classes, however, leads to a generic classification of families of functions that is stable to perturbations and hence will be very useful in our study of PESs. A third reason for the importance of stability stems from the applications of singularity theory to the experimental sciences. It is customary to insist on the repeatability of experiments. Thus similar results are expected under similar conditions, but since the conditions under which an experiment takes place can never be reproduced exactly, the results must be invariant under small perturbations and hence must be stable to those perturbations. Thus we see it is reasonable to require that the mathematical model of a physical process have the property of structural stability. In order to define this concept of stability, we first need a notion of the equivalence between objects. This is usually given by defining two objects to be equivalent if one can be transformed into the other by diffeomorphisms of the underlying space in which the objects are defined. For the specific case when the objects are PESs, these diffeomorphisms are coordinate transformations and will be required to be smooth, that is, differentiable to all orders, and invertible. This invertibility is a requirement of the Directed Trees method and forms an important reason for employing GAs in the numerical applications of the DT method.

The mathematical branch of differential topology called catastrophe theory forms the foundation for the DT method. In its usual form, catastrophe theory is merely a classification of degenerate singularities of mappings, the techniques of which use singularity theory and unfolding theory extensively along with a very important simplifying observation made by Thom which has come to be called the Splitting Theorem. In this paper, we wish only to emphasize the fundamental concepts behind the Classification Theorem thus providing a heuristic justification for its use in the study of PESs. In the process we describe the functional relationships between the PES and its canonical form, which we call the Directed Trees Surface (DTS). We do not provide rigorous statements nor proofs of any of the theorems of differential topology, but, more importantly, we hope to provide an intuitive description of the fundamental concepts behind these theorems. In order to describe these results, we employ the terminology of differential topology, and thus below we provide the basic definitions necessary for comprehension and discussion of the DT method. A glossary of topological terms and notation used, sometimes without comment, in this paper is provided in the Appendix. Since our main interest in this paper is the local properties of potential energy functions, we begin by recalling some preliminary definitions of local properties. If two functions agree on some neighborhood of a point, then all of their derivatives at that point are the same. Thus, if we are interested in trying to deduce the local behavior of a function from information about its derivatives at a point, we do not need to be concerned with the nature of the function away from that point but may only be concerned with the function on some neighborhood at this point. This leads to the concept of a *germ* of a function. Let L be the set of all continuous functions from the Euclidean space $\Re^n$ to $\Re$ defined in a neighborhood of the origin. We say that two such functions, $f, g \in L$ determine the same *germ* if they agree in some neighborhood of the origin, so that a germ of a function is an equivalence class of functions. Since this theory is entirely local, we may speak of the values of a germ $\tilde{f}$ and we write $\tilde{f}(x)$ for $x \in \Re^n$, although it would be more correct to choose a representative f from the equivalence class $\tilde{f}$. A germ $\tilde{f}$ at x is *smooth* if it has a representative which is smooth in the neighborhood of x. Because germs and functions behave similarly, we often use f and $\tilde{f}$ interchangeably to represent a germ. Only where confusion may result will we distinguish a germ from one of its representatives. We may also talk of germs at points of $\Re^n$ different from the origin. A germ is thus defined by a local mapping from some point of origin. If two smooth functions have the same germ at a point, then their Taylor expansions at that point are identical. We may, without loss of generality, take the origin of a germ to be the origin of $\Re^n$. The set of all germs from $\Re^n$ to $\Re$ forms an algebra.[4] This convenient fact allows us to study the germs of maps with powerful algebraic techniques that ultimately lead to algebraic algorithms for the topological study of arbitrary PESs.[5]

Fundamental to many applications of applied mathematics is the technique of representing a function by a finite number of terms in its Taylor expansion. For quantitative calculations, it is necessary to make some estimate for the size of the remainder term after truncation of the series. Sometimes we are not interested so much in the size of the remainder term as in whether, by a suitable change in coordinates near x, the remainder term can be removed completely. In this case, the function is, in a very precise sense, equal to its truncated Taylor series in the new coordinates. The ability of transforming away the higher-order terms of a Taylor series expansion is formalized in the notion of *determinacy*. Before defining determinacy, we first introduce some additional nomenclature. The Taylor series of f at x which is truncated after terms of degree p is referred to as the *p-jet* of f at x, denoted by $j^p f(x)$. We now define what we mean by the local equivalence of germs. Two germs, f, g with $f(x) = g(x)$, are *equivalent* if there exists local C^∞-diffeomorphisms $\psi: \Re^n \to \Re^n$ and $\phi: \Re \to \Re$ such that $g = \phi(f(\psi(x)))$. Thus, by suitable C^∞ changes of local coordinates, the germ f can be transformed into the germ g. We now note why the coordinate changes must be invertible. Neglecting a constant, the two functions are equal on some neighborhood of a point, and we have expressed f as a function of x, that is, $f(\psi(x))$. In addition, we would like to be able to express g as a function of the coordinates for f, that is, $g(x(\psi))$. This requires us to invert the ψ coordinate transformation: $x = x(\psi)$. As we stated earlier, this invertibility criterion becomes an important reason for choosing GAs to solve the systems of nonlinear equations that arise from the DT method. With this, we may now formulate the definition of determinacy. The *p-jet* ς at x is *p-determined* if any two germs at x having ς as their *p-jet* are equivalent.

If we are studying a C^∞-function f, we may understand its local behavior by expanding f in a truncated Taylor series, ignoring all of the higher-order terms of degree greater than p. We can be sure that nothing essential has been thrown away if we know that f is *p-determined*. Stated more precisely, we may study the topological behavior of a *p-determined* germ f by studying its *p-jet* $j^p f$. One might think at first that no germs are *p-determined* for finite p. As an example of this, consider the germ of f at the origin of $\Re^2$ given by $f(x, y) = x^2$. This is not *p-determined* for any p, since the following function, which has the same *p-jet* as f, $g(x, y) = x^2 + y^{2p}$, is 0 at the origin and positive elsewhere, whereas f is also 0 along the y-axis. However, if f were a function of x alone, $f(x) = x^2$, then f would be 2-determined. We thus see that the determinacy of f depends not only on its form but also on the domain over which it acts. Since we have noted that if a function is *p-determined*, its topological behavior may be understood by studying its *p-jet*, then we may now ask the following question: Are there methods for deciding whether or not a given *p-jet* is determined? We answer this question in the affirmative, and in a later paper we describe an algorithm based on work by Mather for calculating the determinacy of *p-jets*.[5] In Section 4 of this paper, which describes the fitting of DTSs to PESs, we provide examples for (i) the DTS behavior for cases in which the proper *p-jet* is chosen for f, (ii) the behavior for cases in which the chosen *p-jet* has p less than the determinacy of f, and (iii) the behavior of $j^p f$ in which p is greater than the determinacy of f.

Below we summarize the four basic and interrelated concepts of singularity theory: (i) stability, (ii) genericity, (iii) reduction, and (iv) unfolding of singularities. To describe what is meant by stability, consider the map $f: \Re \to \Re$ given by $f(x) = x^2$. This map is stable, since we may perturb the graph of this map slightly and the topological picture of its graph remains the same. That is, consider the perturbed map $g: \Re \to \Re$, $g(x) = x^2 + \epsilon x$ with $\epsilon \neq 0$. This perturbed function, g, still has a single critical point just as f does, and can be shown to be just a reparametrization of f. Thus we hope to characterize and classify stable maps since if we perturb these, we can still predict their topological behavior.

Since our goal is to provide a mathematical model for classifying and calculating PESs, one might ask whether there are enough stable maps to be worthwhile in this endeavor. That is, can any arbitrary PES be approximated by a stable map? This is the question of the genericity of stable maps, *i.e.*, whether the set of all stable maps is open and dense in the set of all maps. If it is, then any map is arbitrarily "close" to a stable map and may be represented by combinations of stable maps. It thus makes sense to study the properties of stable maps since these properties will then be pertinent to any arbitrary PES.

Reduction refers to the often employed technique of splitting a problem into two components: one component whose behavior is simple and known, and a second component whose behavior is unknown and hence more interesting and whose behavior we would like to study. This is typical in most physical models in which there are many variables whose functional behavior is assumed to be simple, for example, harmonic. These variables are usually "factored out" of the overall model for the physical phenomenon since the behavior of the system over these variables is known. The Splitting Theorem provides a justification for this reductionism.

René Thom introduced the basic notion of the unfolding of an unstable map in order to provide stability for a family of maps. To see what this means, let us consider the following example to which we will often return for illustrating new topological concepts. Let $f: \Re \to \Re$ be given by $f(x) = x^3$. This map is unstable at zero, since if we perturb f by ϵx, where ϵ is small, the perturbed map $g(x) = x^3 + \epsilon x$ assumes different critical behaviors for $\epsilon < 0$ and $\epsilon > 0$. There are two critical points, a minimum and a maximum, in a small neighborhood of zero when $\epsilon < 0$, but for $\epsilon > 0$ there are no critical points. The family of maps $F(x, \epsilon) = g(x)$ is, however, stable. Thus F includes not only f, but also all possible ways of perturbing f. The map F is said to be an universal unfolding of f. It is very important that the unfolding F includes all possible ways of perturbing f. To be more specific, consider perturbing f by the term δx^2, where δ is arbitrarily small but not zero. The map $h(x) = x^3 + \delta x^2$ assumes the same critical behavior for all $\delta \neq 0$, that is, $h(x)$ has one maximum and one minimum. Thus for $\epsilon < 0$, $g(x)$ has the same critical behavior as $h(x)$, and it can be shown that g and h are "equivalent" for $\epsilon < 0$ and $\delta \neq 0$. (The precise meaning of "equivalent" is described in the Glossary.) On the other hand, there is no δ for which $h(x)$ lacks critical points, thus $h(x)$ is not equivalent to $g(x)$ when $\epsilon > 0$. Therefore h is not capable of describing all possible perturbations of f, since it is unable to provide g with $\epsilon > 0$. The unfolding g is, however, capable of describing all possible perturbations of f. Our discussion so far does not indicate how we know this fact; it is a rather deep result of singularity theory stemming from results based on the early insights of Thom. The crux of singularity theory is how to unfold the "interesting" component of a given model into a stable mapping with the least number of parameters, such as the ϵ from above.

3.1. *Theorems from Topology*

Several principal theorems of differential topology concern the effects that critical points have on the geometrical shape of manifolds. Since each has been carefully proven and thoroughly investigated in the literature, we only include here an informal statement of these theorems and a few of the results derivable from them. We emphasize that these theorems are closely related to each other; their differences entail the stepwise removal of some of the assumptions upon which the first theorem is based.

The first of these theorems is borrowed from elementary calculus: the Implicit Function Theorem.[6] This theorem controls the behavior of a surface at regular points, that is, at points which are not critical points. Excluding the overall translational and rotational coordinates of a molecule, the critical points of potential energy surfaces are isolated.[7] Thus almost all points of a PES are regular points and hence the implicit function theorem describes the local behavior of almost all of a PES. Qualitatively speaking, the Implicit Function Theorem states that at a noncritical point of a potential function, the coordinate axes may be rotated so that one of the axes aligns with the gradient of the potential at that point. Then the function is represented as $f(\vec{x}') = x'_1$ where $\vec{x}'$ are the new coordinates. This is intuitively obvious by considering the gradient to be a "force vector", then the coordinate axes may be rotated so that one axis is colinear with the force, which may then be described as a linear function of this one coordinate. In analogy to our one-dimensional example of the control which critical points have on the possible shape of a function, we find that the overall shape of a PES depends upon the positioning and type of its critical points. The Morse Theorem, which is sometimes called the Morse Lemma in the literature,[8] and its corollaries describe how nondegenerate critical points both control the shape of a surface and determine the relationship between an approximately measured function and the stable mathematical model which is used to descibe that physical process. In particular, through the elimination of the assumption that the gradient is nonzero at a point, we find around nondegenerate critical points, a new coordinate system so that a potential may be represented as the sum of squared terms of the coordinates with no higher-order terms, no linear terms, and no quadratic cross terms. Thus the function has the form $f = \sum x'^2_i$ and is termed a Morse function. Corollaries of the Morse Theorem say that Morse functions are stable and this stability is a generic property. Lastly, we discuss degenerate critical points and their influence on the possible configurations of nondegenerate points. By eliminating the assumption of the nonsingular Hessian matrix at a critical point of the surface, the Gromell-Meyer Splitting Theorem says that the function may be split into two components, one is a Morse function, F_M, and the other is non-Morse function, F_{NM}. The non-Morse component cannot be represented as quadratic terms and does not involve any of the coordinates of the Morse component! The Arnol'd-Thom Classification Theorem[9] categorizes all of these non-Morse functions into families, provides canonical forms for them, and describes the interrelations among the various families.

The ramifications of the Arnol'd-Thom theorem cannot be overestimated. If a function, $F(\vec{x}; \vec{\rho})$, having a non-Morse critical point at $(\vec{x}_c; \vec{\rho}_c)$ is perturbed. The perturbed function, $F'(\vec{x}; \vec{\rho})$, through diffeomorphisms $\vec{\chi}$ and $\vec{\rho}$, is obtained from F by perturbing the Morse part and the non-Morse part separately. Perturbation of the former does not change its qualitative critical behavior, while perturbation of the later does. Thus one can "forget" about the coordinates involved in the Morse function, while concentrating on the subspace spanned by the variables of F_{NM}. The theorem classifies all possible types of perturbed functions in this subspace. Corollaries also establish the stability and genericity of the universal unfoldings of the Classification Theorem.

3.2. *Potential Functions and their Canonical Forms*

In this section we want not only to discuss the connection between arbitrary potential functions and their canonical forms provided in a separate paper,[1] but also to demonstrate the quantitative relationships that exist between the critical points, gradients, and curvatures of the potential function with the corresponding expressions that exist for the canonical forms. In order to define the extent of the applications of these canonical forms we begin with a brief exposition of Thom's method[10] for modeling a physical system.

First, suppose the physical system we wish to model has n distinct properties to which n definite real values may be assigned. We define an n-dimensional Euclidean space, $\Re^n$, which parametrizes these various physical variables. Each point in $\Re^n$ represents a particular state for the physical system. If $\vec{x}, \vec{x} \in \Re^n$, is such a point, then the coordinates of $\vec{x}$, $(x_1, \ldots, x_n)$, are called the state variables. Let $\mathcal{X} \subset \Re^n$ be the set of all possible states of the physical system. The particular state, $\vec{x} \in \mathcal{X}$, which describes the system, is determined by a rule which usually depends on a multidimensional parameter represented by $\vec{\rho}$, $\vec{\rho} = (\rho_1, \ldots, \rho_k) \in \Re^k$. For most physical systems this rule is often specified as a flow associated with a smooth vector field, say, $\mathcal{Y}$. This flow, or trajectory, on $\mathcal{Y}$ usually determines the attractor set of $\mathcal{Y}$. Sometimes the rule is specified so the flow "chooses" a particular attractor on $\mathcal{Y}$ with the "largest" basin. At other times the rule may only specify that the attractor be a stable one. Since very little is known mathematically about the attractors of arbitrary vector fields, catastrophe theory has little to say about this general model. If, however, the vector field is further restricted to be one generated by the gradient of a given smooth function, say V, then Thom's theory becomes very useful in the study of the physical model. In other words, if $\mathcal{Y} = -\mathbf{D}V(\vec{x}; \vec{\rho})$ where V is considered a family of potential functions on $\Re^n \otimes \Re^k$, the attractors of $\mathcal{Y}$ are just the local minima of

214

$V(\bar{x};\bar{\rho})$. In terms of a potential function, the rule S again may have several forms. For instance, S may choose the global minimum of V, or it may require only that the state of the system corresond to one of the local minima of V. The specific details of the method which S uses to move $\bar{x}$ to the attractors of $\mathcal{Y}$ determines the dynamics of the trajectory of $\bar{x}$ in $\mathcal{X}$. Various choices for S may correspond to tunneling through barriers on V, to steepest descent paths on V, or to "bouncing" over small barriers by means of thermodynamic fluctuations.

3.3. *Relationships between Potential Functions and their Unfoldings*

In order to examine a specific example, let us suppose that $\mathcal{Y}$ is a gradient vector field: $\mathcal{Y} = -\mathbf{D}V(\bar{x};\bar{\rho})$, where $V(\bar{x};\bar{\rho}): \Re^n \otimes \Re^k \to \Re$ is a smooth potential function of the state variables, $\bar{x}$, and depends upon a parameter $\bar{\rho}$. The attractor set of $\mathcal{Y}$ is then specified as a set of stable minima of V. The critical points of V, defined by $\mathbf{D}V = \bar{0}$, form a manifold, $\mathcal{X}_V$, where $\mathcal{X}_V \subset \Re^{n+k}$, which includes the stable minima. Choosing a point, $(\bar{x}_0; \bar{\rho}_0) \in \Re^{n-k}$, of $\mathcal{X}_V$, Thom's classification theorem tells us that in some neighborhood of $(\bar{x}_0; \bar{\rho}_0)$, V is equal to the sum of a universal unfolding, U_i, of one of the germ functions, G_i, and a quadratic form Q. $Q = \sum_{m=j+1}^{n} x_m^2$ for $k \leq 6$ and $j = 1$ or 2.[9] More formally, if $\mathcal{N}_x \subset \Re^n$ is a neighborhood of $\bar{x}_0$ and $\mathcal{N}_\rho \subset \Re^k$ is a neighborhood of $\bar{\rho}_0$, then $V: \mathcal{N}_x \otimes \mathcal{N}_\rho \to \Re$ is equivalent to $F_i(\bar{x};\bar{\rho}) = G_i(\bar{x}_{1,j}) + P_i(\bar{x}_{1,j};\bar{\rho}) + Q(\bar{x}_{j+1,n}) = U_i(\bar{x}_{1,j};\bar{\rho}) + Q(\bar{x}_{j+1,n})$ for some finite i with $\bar{x}_{1,j}$ denoting the first j coordinates of $\bar{x}$ while $\bar{x}_{j+1,n}$ denotes the last $n - j$ coordinates. This means that there exist diffeomorphisms $\bar{\chi}: \mathcal{N}_x \otimes \mathcal{N}_\rho \to \mathcal{N}_x$, and $\alpha: \mathcal{N}_\rho \to \Re$ such that, for any $(\bar{x};\bar{\rho}) \in \mathcal{N}_x \otimes \mathcal{N}_\rho$, we have

$$V(\bar{x};\bar{\rho}) = F_i(\bar{\chi}(\bar{x};\bar{\rho}); \bar{\rho}(\bar{\rho})) + \alpha(\bar{\rho}) \tag{3}$$

This equation allows us to quantitatively relate the critical points, gradients, and curvatures of V and F_i. Application of the chain rule for derivatives of vector fields to equation (3) provides an expression for the gradient of V:

$$\mathbf{D}V(\bar{x};\bar{\rho}) = \mathbf{D}F_i(\bar{\chi};\bar{\rho})\mathbf{D}\bar{\chi}(\bar{x};\bar{\rho}) \tag{4}$$

where $\mathbf{D}$ denotes the partial derivative operator with respect to the coördinates of the function or operator which follows it. In order to determine the Hessian of V, $\mathbf{H}V$, we carefully reapply the chain rule to (4) to yield:

$$\mathbf{H}V(\bar{x}) = \mathbf{D}^t \bar{\chi}(\bar{x}) \bullet \mathbf{H}F_i(\bar{\chi}) \bullet \mathbf{D}\bar{\chi}(\bar{x}) + \sum_{k=1}^{n} \mathbf{D}_k F_i(\bar{\chi})\mathbf{H}\chi_k(\bar{x}) \tag{5}$$

where $\mathbf{D}^t$ is the transpose of $\mathbf{D}$. We now have expressions equating not only V and F_i, (3), but also their gradients, (4), and Hessians, (5). Through these systems of nonlinear equations the unfolding parameters and diffeomorphisms may be calculated.

As Connor[11] has pointed out in a different context, the diffeomorphism and parameters of an unfolding may be calculated *via* the solution of the nonlinear system of equations which arises from the correspondence between the critical points of the unfolding and those of the experimental function. For PESs, however, the critical points are usually not known *a priori*, and thus this is not a viable procedure. Extensions of this method are reasonable though. For instance, the DTS and PES must correspond within a neighborhood of any point. Thus, a similar system of nonlinear equations may be derived, for points within some neighborhood of a particular point, whose solution yields the parameters and diffeomorphism. Alternatively, at a single point the function and all of its derivatives must coincide with those of the DTS. Therefore, since *ab initio* quantum calculations now provide analytic first and second derivatives, it is reasonable to employ this information to help calculate the DTS parameters and diffeomorphism. Thus, the calculation of a single point on the PES with its first and second derivatives may be employed to determine a first approximation to the parameters and diffeomorphism. Thus, from a single point, we may be able to specify to which unfolding within a given family the particular PES belongs. Since there are canonical forms for the DTS, we also have canonical forms for its critical points, in particular, its saddle points.[12] Therefore, one might next move over to the DTS saddle point and perform another quantum mechanical calculation there. Of course, this point will not correspond to the PES saddle point, but since locally the diffeomorphism is approximately the identity function, it will be close to the PES saddle point. The additional information obtained at this new point may then be used to calculate a second approximation for the parameters and diffeomorphism. Thus, with each new point, better parameters are calculated so that the DTS better fits the PES. In the next section, we perform sample calculations on the one-dimensional unfolding families, the A_i families.

Section: 4. DT Method for fitting a PES *via* the Genetic Algorithm

As we discussed in the last section, the problem of fitting a DTS to a PES is one of finding a solution to a nonlinear system of equations. The DT method allows for a flexible choice for the form of the optimization function. We have considered both weighed least squares as well as absolute value evaluation functions. In particular, in the follwoing examples we have employed the experimental and evaluation functions provided below:

$$\text{Experimental function:} \quad f(x) = \alpha x^6 + x^3 + 3x^2 \ , \ \alpha = -0.05$$

$$A_2 \text{ Unfolding:} \quad F(\chi) = \chi^3 + \rho_1 \chi + \rho_0$$

$$\text{Diffeomorphism:} \quad \chi(x) = c_0 + c_1 x + c_2 x^2$$

$$\text{Evaluation functions:} \quad R = \sum_i w_i \left| F(\chi_i) - f(x_i) \right|$$

$$R_1 = \sum_i w_i \left| \frac{\partial F(\chi_i)}{\partial x} - \frac{\partial f(x_i)}{\partial x} \right| \tag{6}$$

$$R_2 = \sum_i w_i \left| \frac{\partial^2 F(\chi_i)}{\partial x^2} - \frac{\partial^2 f(x_i)}{\partial x^2} \right|$$

$$R_3 = r R + r_1 R_1 + r_2 R_2$$

$$\text{where} \quad \{r, r_1, r_2, w_i\} \text{ are weighting factors.}$$

The standard numerical methods for solving nonlinear systems often involve algorithms of the Newton-Raphson type.[13] As we mentioned earlier, the coordinate transformation must be a diffeomorphism, and hence, invertible. Empirically, we found that when employing a Newton-Raphson algorithm for solving these nonlinear systems, the calculated coordinate transformations often did not satisfy the invertibility criterion. Therefore we resorted to constrained optimization techniques. Several methods, including the Box complex algorithm,[14] and standard least squares procedures,[15] have been successfully used to solve these nonlinear equations. Typically, the constained methods were very slow to converge to a minimum and thus required a significant increase in computational time. Since the evaluation functions involved the differences between values for the experimental PES and its DTS, they were froth with shallow local minima. Thus, for some problems, these methods did not converge to the global minimum of the evaluation functions. In addition, the constrained optimizations often tended to remain close to their constraint boundaries, resulting in the optimizations becoming stuck in local minima. These considerations led us to try other function optimizers. Besides these classical techniques, genetic adaptive algorithms (GAs) also may be employed to solve these systems. GAs are based on an observation originally made by Holland[16] that living organisms are very efficient at adapting to their environs. Implicit in a genetic adaptive search is an immense amount of parallel calculation, and empirical studies indicate that GAs will often outperform the usual numerical techniques.[17] We do not discuss the working of GAs here, but rather refer the reader to literature references.[18]

Several features illustrated in the following fitting examples are of importance and we mention them here: (i) We show that the coordinate transformation employed by the DT Method is required to be a diffeomorphism. If the coordinate transformation calculated *via* the DT method is not a diffeomorphism, then the chosen determinacy of the PES is too low and a higher-order unfolding family is needed in order to accurately fit the PES. (ii) Also illustrated is the fact that the diffeomorphism may include terms which have asymptotic behavior, for example, exponential terms. In this case, the asymptotic behavior of the surface may be reproduced by including comparable behavior in the diffeomorphism. (iii) "Bumps" or "shoulders" on surfaces that do not form critical points still reflect the fact that they stem from the annihilation of critical points of a germ function. Thus any bump or shoulder on a surface means that a higher order unfolding family will be required in order to accurately reproduce them. (iv) Also depicted in these examples is the DT Method for fitting a 2-dimensional potential energy surface. Our example 2-D surfaces have one "interesting" coordinate, that is, one coordinate which is not harmonic, and one coordinate which is harmonic.

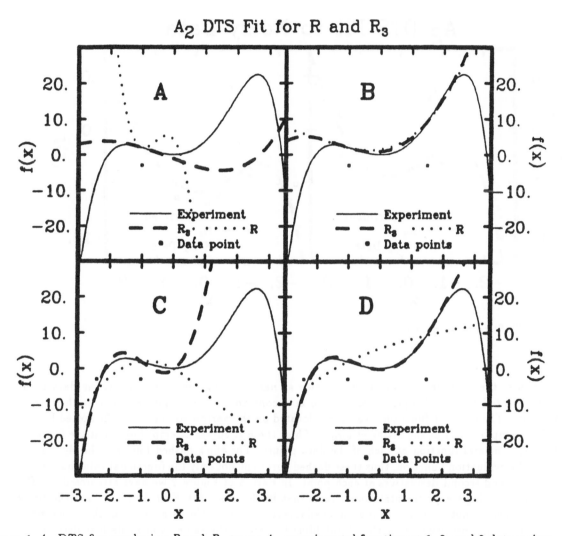

Figure 1 A_2 DTS fits employing R and R_3 to an A_5 experimental function at 1, 2, and 3 data points.

In Figure 1, we illustrate the Directed Trees fitting procedure by employing the genetic algorithm for fitting the A_2 unfolding family to an experimental function belonging to the A_5 family. We choose this experimental function to exemplify several features of the DT method. In particular, the value of the coefficient of the x^6 term was chosen in order to generate a third critical point on the experimental surface within the coordinate interval $-3 \leq x \leq 3$. We choose this interval so that the local nature of the fitting procedure for the A_2 unfolding may be demonstrated. In conjunction with this local aspect of the A_2 DTS on the $[-3,3]$ interval, however, we would like to point out that all three critical points, and hence the experimental function itself, may be accurately represented with the A_3^- unfolding family. Even though the highest-order term of the A_3 germ function is fourth-order, its unfoldings may have three critical points and thus the three critical points of this A_5 PES may be accurately reproduced on the interval $[-3,3]$. We have successfully fit an A_3^- DTS to all three singularities of this A_5 PES (The A_3^+ unfolding family does not have the proper local topology and consequently it cannot accurately reproduce this PES. When an A_3^+ DTS fit is attempted, either the fitting is very poor or the calculated coordinate transformation is not a diffeomorphism.) This example also demonstrates the usage of the DTS to help choose new positions for further calculations and the employment of the first and second derivatives in addtion to the functional values at the data points.

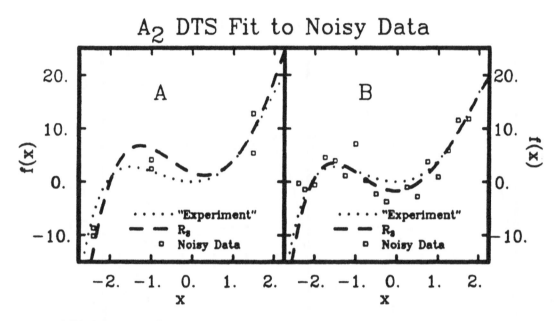

Figure 2 A_2 DTS fit to noisy data points.

In this figure, the experimental function is drawn as narrow solid lines. For clarity, the data points, which are represented as "solid" squares, are drawn at a constant 'y' coordinate and not at their proper functional values. Their proper functional values are located on the narrow solid curve. The dotted lines are the A_2 DTS fits employing R as the fitting criterion. Thus, these R curves attempt only to fit the functional value of the experimental function at each of the data points. The thick dashed lines are the A_2 DTS fits employing the R_3 evaluation fiunction, thus these dashed curves fit not only the functional value but also the values of the first and second derivatives at each point. In Part A of Figure 2 we have attempted the DTS fit employing only a single experimental point. Note that in this case, the R fit does not have the proper local topology. There is not enough information available to determine the local shape of the experimental function, and it is only fortuitous that the R unfolding has about the same value of its first derivative as the experimental function. On the other hand, the R_3 DTS fit does have the proper local topology but its critical points are far removed from the corresponding experimental minimum and maximum. In Parts B and C of this figure, we employ two experimental data points for fitting the DTSs. In Part B, the chosen data points include the single point from Part A plus an additional point at the minimum of the DTS surface calculated in Part A. We thus have used the approximate DTS surface of Part A to choose where the next calculation should be performed. The new information from the second datum point is then used to refine the DTS. In Part C, we use the same datum point as in Part A as well as the maximum point of the DTS in A. These refined DTS curves in Parts B and C now provide a more accurate estimates of the minimum and maximum of the experimental function. We use three data points for Part D, the original point from A as well as Part A's DTS's minimum and maximum points. Note that the R DTS fit to the three points does not have the proper topology of the experimental function. The R_3 DTS, however, is a very accurate fit within the neighborhoods surrounding the maximum and minimum of the experimental function. Note, however, that the A_2 DTS is unable to fit the second, rightmost, maximum of the experimental function. This is because this third critical point generated by the sixth-order term in the experimental function cannot be represented within the A_2 unfolding family, which has, at most, two critical points. A higher-order family would have to be chosen in order to fit this maximum value. In particular, the A_3 family would be capable of fitting both of the maxima and the minimum on this experimental function. One does not have to use an A_5 unfolding for this experimental function even though it contains a sixth-order term.

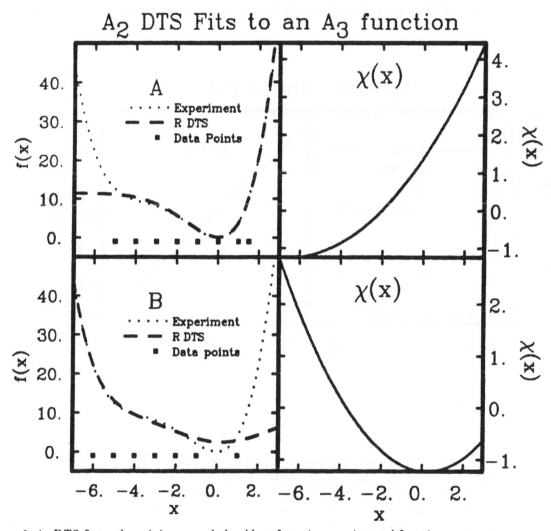

Figure 3 A_2 DTS fit to the minimum and shoulder of an A_3 experimental function.

In the previous figure, the experimental function was assumed to be known exactly. This is usually not the case. Typically, there are random errors in the potential energy at each datum point on a PES. These random fluctuations stem from round-off errors in calculations, approximate wave functions, numerical integration inaccuracies, or experimental random fluctuations. Also, as previously noted, the evaluation functions have many local minima which often appear to be similar to random fluctuations. To show that the DT method in conjunction with the GA optimizer does not require exact data, we return to the experimental function of Figure 2 in the next figure and add rather severe random fluctuations to the functional of values as well as its first and second derivatives. The GA optimizer is very efficient at avoiding local minima and consequently works well for noisy PESs. Part A of this figure has six data points to which noise has been added. (The "open" squares representing the data points in this figure now reside at their proper functional values.) Note that the best A_2 DTS fit employing R_3 to these data does not accurately repeat the "exact" experimental function, that is, the function without the random fluctuations which is drawn as a dashed line. In fact, it might appear as if the DTS does not even accurately fit the two data points at $x = -1$. It must be realized that the functional value of the data points is all that is being plotted in this diagram. The R_3 evaluation function, however, includes the first and second derivatives as criteria for a fit. Thus for a small number of data points, the random fluctuations in the first and second derivatives need not cancel and thus the DTS need not accurately fit the two functional values at $x = -1$. In Part B, we have added additional data points. Here, the DTS fairly accurately fits the "exact" experimental function. This figure illustrates that the Directed Trees method coupled with the genetic algorithm are easily applied to fitting DTSs to noisy PESs.

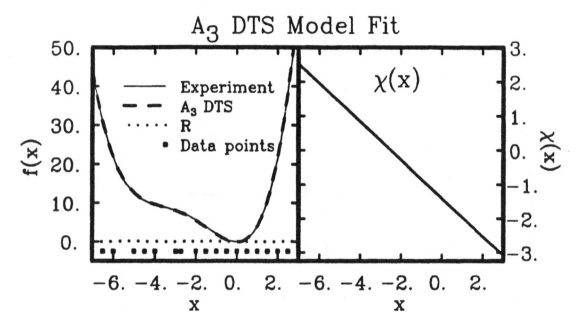

Figure 4 A_3 DTS fit to the A_3 experimental function.

One particular advantage of employing the genetic algorithm for fitting DTSs to PESs is that it is easy to require that the calculated coordinate change remain a diffeomorphism. In the next figure, we see not only a new experimental function as well as its DTS fits, but in addition, plots of the corresponding diffeomorphisms, $\chi(x)$, for the DTSs. Note that in Part A, we have chosen data points surrounding the minimum of the experimental function at $x = 0$. This experimental function has only a single minimum, but it does have a shoulder at around $x = -3$. Even though this shoulder is not a new critical point, it stems from the annihilation of a saddle and a minimum of the A_3 family of functions. Hence our A_2 DTS cannot fit this experimental function exactly. It is capable of fitting either the minimum as illustrated in Part A or the shoulder as illustrated in Part B. In addition, Part A also illustrates the possibility of asymptotic behavior being included in the diffeomorphism, then the DTS is capable of fitting the asymptotic behavior on a PES. In fact, instead of expanding the diffeomorphism as a Taylor series, as we have done here, it could easily be expanded as a sum of exponential terms whose asymptotic behaviors are then imparted to the DTS. Note that, as the diffeomorphism levels off for $x < -5$, the DTS also becomes asymptotically level. Part B of this diagram contains a warning, however. The function $\chi(x)$ is not a diffeomorphism over the entire interval, $-7 \leq x \leq 3$, and hence, the assumptions necessary for application of the Arnol'd-Thom

Classification Theorem are not satified over this interval. In fact, the critical point of $\chi(x)$ leads to an additional critical of the DTS at about 0. This critical point of $\chi(x)$ was induced by attempting to fit the A_2 unfolding family to "three" critical points: the one actual minimum of the surface and the annihilated saddle and minimum which generates the shoulder region. If the datum point at $x = 1$ is removed, then $\chi(x)$ remains a diffeomorphism and the DTS accurately fits the shoulder of the experimental function. This example reveals an advantage of the genetic algorithm over many of the nonlinear Newton-like optimization schemes. Unlike the Newton methods which require an initial guess and can become "stuck" in local minima, the genetic algorithm only requires starting intervals for its parameter values. This, by the way, allows one to assure that the coordinate transformation χ remains a diffeomorphism by means of controlling the ranges over which the parameter values may vary. In addition to the fact that parameter intervals are a much less restrictive initial condition than having to guess a starting parameter solution, one may also easily specify the resolution at which each individual parameter is calculated. Thus individual parameters may all be optimized at differing resolutions. If χ is not a diffeomorphism after fitting a DTS to a PES, then this is a tipoff that the chosen fitting family is too small and does not contain enough critical points necessary for fitting the surface. Thus one should choose a higher-order family for fitting this surface. In particular, the next figure, Figure 4, illustrates that if we choose the A_3 unfolding family to fit this experimental function, then both the shoulder and the minimum may be accurately fit. Since this experimental function is 3-determined and we are employing the A_3 unfolding family, the diffeomorphism is a linear function with no critical points.

We next consider the Directed Trees method applied to an experimental function which has more than one dimension. We choose an experimental function which has one nonharmonic coordinate and one harmonic coordinate. This PES is representative of isomerization reactions. It is an important trial case because of the recent interest in quasi-periodic versus chaotic trajectories on similar two-dimensional surfaces.[19] Also a similar surface was also chosen by Fukui[20] to illustrate the intrinsic reaction coordinate method. Contour levels of this function are drawn in Parts A and C of the following figure. There are several things to note about the experimental function. First of all, there are two minima and one saddle point. Neither of the minima are located at special points, such as the origin. Also, a line drawn between the two minima is not parallel to either of the coordinate axes. The DT method, though, is capable of "rotating" the DTS coordinate axes so that it can accurately represent the experimental surface. In Part B of this figure, we have chosen the A_3 family for fitting this function. Note that the corresponding contour levels in all Parts of this diagram are drawn employing the same type of line, whether that be solid, dashed, dash-dotted, or dotted. The "stars" (*) in Parts A and C locate the data points used in the calculations for Parts B and D, respectively. The A_3 DTS of Part B very accurately fits the experimental function.

You might ask what would happen if one were to choose a family which can display more critical points than the experimental function contains. This is illustrated in Part D of this Figure. In this case, the A_4 unfolding family was chosen to fit the same experimental function as provided in Part A. Note than in Part D, the DTS accurately fit both minima and the saddle point of the experimental function. In addition, however, there is a new saddle point appearing around the point $(2.1, 0.2)$. This new saddle point stems from the fact that the A_4 family can display four critical points. It is worth noting, however, that in the region surrounding the data points, the A_4 DTS accurately fits the experimental function. The new, extraneous, saddle point of the DTS lies outside the local neighborhood of the data points employed to fit this PES. This example of employing the A_4 unfolding mightlead one to consider always employing a high-order unfolding family to fit all PESs. One finds, however, from the practical viewpoint of calculating the fitting parameters, that a properly chosen unfolding family (one whose determinacy and local topology is the same as the experimental PES) will greatly reduce the amount of calculation and hence provide an easily calculated fit to the PES. This is because the DTS has the proper number of critical points to reproduce the topology of the surface: data is not required to suppress extraneous critical points of the unfolding. Thus there is an optimum unfolding family, from a calculation standpoint, for each PES. It is true that the higher-order family, assuming it contains the lower-order family as a subfamily, will provide an unfolding which repeats the topology of its lower-order subfamily. It is this subfamily, however, that should be chosen as the unfolding family for the original fitting procedure.

As our last example of a 2-dimensional fitting to a 2-dimensional PES, we choose the same "exact"

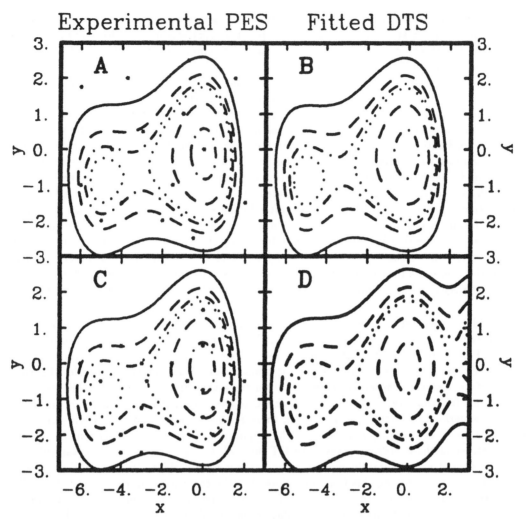

Experimental PES   Fitted DTS

Figure 5 A_3 and A_4 2-dimensional DTSs fit to an experimental function. Contour lines are drawn at energies of 15, 10, 8, 7, 4, and 1 in all parts of this figure.

experimental function, but add in random fluctuations to the experimental values and its first and second derivatives. For this example, we also employ the R_3 evaluation function in determining the unfolding and diffeomorphism parameters. Note that in Figure 6 the A_3 DTS has the same critical behavior as the exerimental PES, however, it is not as accurate of a fit as that shown in Figure 5. This is because the noise included in our functional values is rather extensive. Since it is not possible to see these random fluctuations on a contour plot of the PES, we have drawn a 3-D stereo projection of the experimental PES along with the noisy data points chosen. In this view, the bold circles are the experimental points chosen on the surface while the light crosses are the "exact" experimental values corresponding to the noisy data points.

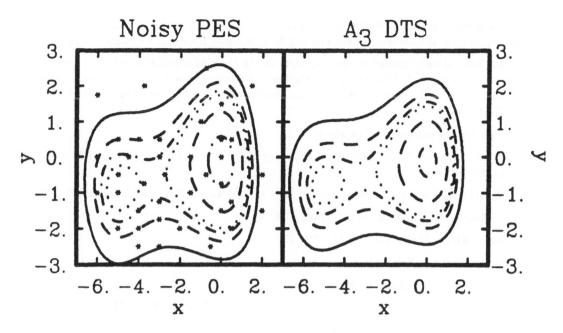

Figure 6 A_3 DTS fit to a noisy A_3 experimental 2-D function.

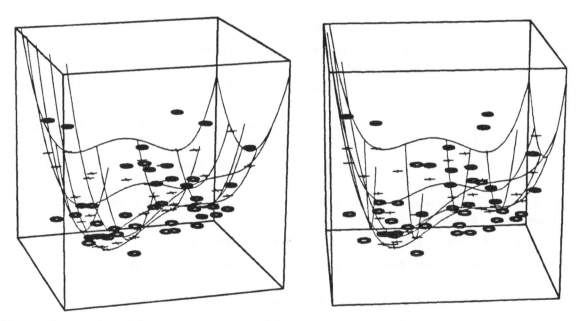

Figure 7 Stereo view of the noisy data points of the A_3 experimental 2-D function. The bold circles are the noisy data points while the thin "+" signs are the corresponding "exact" values.

Section: 5. References

[1]   See the associated paper: "The Directed Trees Method I: Classification of Potential Energy Surfaces" (submitted for publication).

[2]   For typical examples of the simplifications that may arise solely from a classification scheme, see "The Differential Topology of the Directed Trees Method V: Symmetry Invariant Potential Energy Surfaces."

[3]   These concepts have already played important parts in the story of classical mechanics and dynamics, for example, see Arnol'd, V. I. "Mathematical Methods of Classical Mechanics"; Springer-Verlag: New York, 1978, and Arnol'd, V. I.; Avez, A. "Ergotic Problems of Classical Mechanics"; Benjamin: New York, 1968; Ch 1, 3–4.

[4]   See the Glossary for the definition of an algebra.

[5]   See the accompanying paper "The Differential Topology of the Directed Trees Method III: Determinacy and Unfolding Algorithms."

[6]   Rudin, W. "Principles of Mathematical Analysis"; 3rd Ed.; McGraw-Hill: New York, 1976; p 224.

[7]   Mezey, P. G. *Theoret. Chim. Acta (Berl.)* 1981, *58*, 309.

[8a]   Morse, M. *Trans. Amer. Math. Soc.* 1931, *33*, 72.

[8b]   Milnor, J. "Morse Theory"; Princeton Univ. Press: Princeton, New Jersey, 1963. No. 51.

[9]   Arnol'd, V. I. *Russian Math. Surveys* 1974, *29*, 10.

[10]   Thom, R. "Structural Stability and Morphogenesis"; Benjamin: Reading, MA, 1975.

[11]   Connor, J. N. L. *Mol. Phys.* 1976, *31*, 33.

[12]   See the accompanying paper "The Differential Topology of the Directed Trees Method II: Potential Energy Surfaces and Canonical Forms."

[13]   Fletcher, R. "Practical Methods of Optimization, Vol. 1 Unconstained Optimization"; Wiley: New York; 1980; Ch 6.

[14]   Richardson, J.A. *Commun. ACM* 1973, *16*, 487.

[15]   International Mathematical & Statistical Libraries, Inc. Subroutines ZXSSQ, ZSPOW, and ZXCNT from "IMSL Library of Fortran Subroutines"; 9th ed.; IMSL, Inc.: Houston, TX; 1981.

[16]   Holland, J.H. "Adaptation in Natural and Artificial Systems"; Univ. of Michigan Press: Ann Arbor; 1975.

[17]   De Jong, K.A. "Analysis of the Behavior of a Class of Genetic Adaptive Systems"; PhD dissertation, Univ. of Michigan, August, 1975.

[18a]   Bethke, A.D. "Genetic Algorithms as Function Optimizers"; PhD dissertation; Univ. of Michigan, January, 1981.

[18b]   Brindle, A. "Genetic Algorithms for Function Optimization"; C.S. Department Report TR81-2 (PhD dissertation); Univ. of Alberta, 1981.

[18c]   De Jong, K.A. *IEEE Trans.: Systems, Man, and Cybernetics* 1980, *10*, 9.

[18d]   Holland, J.H. In *Prog. in Theor. Biol.*, 1976, *4*, 263. Rosen, R.; Snell, F.M., eds.; Academic Press: New York; 1976.

[18e]   Holland, J.H. "Adaptation in Natural and Artificial Systems"; Univ. of Michigan Press: Ann Arbor; 1975.

[19a]   DeLeon, N.; Berne, B.J. *J. Chem. Phys.* 1981, *75*, 3495.

[19b]   Kariotis, R.; Suhl, H.; Eckmann, J.-P. *Phys. Rev. Lett.* 1985, *54*, 1106.

[20a]   Tachibana, A.; Fukui, K. *Theor. Chim. Acta* 1978, *49*, 321.

[20b]   Tachibana, A.; Fukui, K. *Theor. Chim. Acta* 1979, *51*, 189.

Section: 6. Glossary

The following furnishes brief definitions of a few of the terms from differential topology that we employ in the text of this paper.

(1) **C^m-Diffeomorphism:** If ψ is a C^m-diffeomorphism, then it satisfies the following three criteria:

    (i) ψ is m times differentiable,

    (ii) ψ has an inverse, $\psi^{-1}: \Re^n \to \Re^n$, such that $\psi \circ \psi^{-1} = \psi^{-1} \circ \psi = 1$, and

    (iii) ψ^{-1} is m times differentiable,

where m is either finite, ∞, or ω.

(2) **Equivalence class:** If A is a set and if $\sim$ is an equivalence relation on A, then the *equivalence class* of $a \in A$ is the set $\{x \in A | a \sim x\}$.

(3) **Equivalent:** Two functions, $f: \Re^n \to \Re$ and $g: \Re^n \to \Re$, are *equivalent* at 0 if there exists a diffeomorphism $\bar{\chi}: \Re^n \to \Re^n$ and a constant α such that

$$g(\bar{x}) = f(\bar{\chi}(\bar{x})) + \alpha \tag{7}$$

in a neighborhood of 0. Equivalence of two functions implies that they have the same geometric "shape" and critical behavior. They have corresponding critical points which are of the same type.

(4) **Genericity:** A *generic* property is a property possessed by an open dense subset of the system. This means that a generic property is "typical" for the system, and a complementary subset for which the property does not hold has measure 0. Thus it is "mathematically rare" for a generic property not to hold. Since a generic property holds on a dense subset of the system, then any member of the system, including those not having the generic property, may be approximated arbitrarily closely by elements having the generic property. An example of this is that a function having a degenerate critical point may be approximated by functions having only Morse critical points.

(5) **Germ, Germ-equivalent:** Let T be a topological space and S be any set. Let $f: U \to S$ and $g: V \to S$ be maps with domains U, V open sets in T, and suppose x lies in $U \cap V$. Then f and g are said to be *germ-equivalent* at x if there exists some open neighborhood W of x lying inside $U \cap V$ such that $f = g$ on W. This is an equivalence relation on the set of all maps defined on neighborhoods of x in T and with values in S, and the equivalence classes are called *germs* of maps at x. If S is a topological space also, then we can consider germs of continuous maps. If S and T are normed linear spaces, we can consider germs at x of C^r maps. If two C^∞ maps are germ-equivalent at x, then all their derivatives at x are the same.

(6) **k-determined:** Let $f \in \Re^n$ and let k be a non-negative integer. Then f is *right-determined (right-left determined)* if, for every $g \in \Re^n$ such that $j^k(f) = j^k(g)$, then $f \sim_r g$ $(f \sim_{r,l} g)$.

(7) **Jet, k-jet:** The *k-jet* of a function f, denoted by $j^k(f)$, is the Taylor series expansion of f at x and truncated after the order k terms.

(8) **Neighborhood N:** Given a topological space, (T, τ), a subset $N \subset T$ is a *neighborhood* of a point $t \in T$ is there is a member S of τ with $t \in S \subset N$.

(9) **Regular point:** A point, x, is a *regular point* if x is in the domain of a function, $f: \Re^n \to \Re$, and the gradient of the function at x is not zero.

(10) **Smooth or C^∞:** A function f, $f: \Re^n \to \Re^m$, is called *smooth* at a point, x, if all of its derivitives exist and are continuous at x.

(11) **Stability:** Properties of a mapping which are invariant to perturbations of the map are called *stable* properties, and the collection of maps which possess a particular stable property may be referred to as a *stable class* of maps. In particular, a property is stable provided that whenever $f_0: X \to Y$ possesses a property and $f_t: X \to Y$ is a homotopy of f_0, then, for some $\epsilon > 0$, each f_t with $t < \epsilon$ also possesses the property.

(12) **Structural Stability:** For the single function case, let $f: \Re^n \to \Re$ be a function and $P: \Re^n \otimes \Re^k \to \Re$ be an arbitrary small perturbation. Then f is stable at a point in $\bar{x}_0$ if there exists a diffeomorphism $\bar{\chi} = \bar{\chi}(\bar{x})$ such that the perturbed function, $g = f + p$, in the new coordinate system is equivalent to the unperturbed function, $f(\bar{x}) = g(\bar{\chi}) + \alpha$.

(13) **Topology, Topological Space, Open Sets:** Let T be a set; a topology τ on T is a collection τ of subsets on T which satisfy the following criteria: a family τ of subsets on T is a *topology on* T if

 (i) if $\mathcal{T} \subset \tau$, then $\cup \mathcal{T} \in \tau$,

 (ii) if $\mathcal{T} \subset \tau$ and $\mathcal{T}$ is finite, then $\cap \mathcal{T} \in \tau$,

 (iii) $\emptyset \in \tau$ and $T \in \tau$.

then (T, τ) is a *topological space*, T is its underlying set, and the members τ are called the *open* or τ-*open sets* of (T, τ) of T.

(14) **Unfolding, Versal and Universal:** An *unfolding* of a function, $f(\vec{x})$, is a parametrized smooth family of functions, $F(\vec{x}; \vec{p})$, where $\vec{p} = (p_1, \ldots, p_j)$, whose members are possible perturbations of $f(\vec{x})$. The dimension of $\vec{p}$, j, is called the codimension of the unfolding. Usually *unfolding* also refers to a particular member of the family, $F(\vec{x}; \vec{p})$. An unfolding, G, is a *versal* unfolding if any other unfolding of f may be obtained from G *via* a diffeomorphism. An unfolding, H, is a *universal* unfolding if it is both versal and is of minimum codimension.

AUTHOR INDEX